MARVEL UNIVERSE ®

MARVEL

PETER SANDERSON

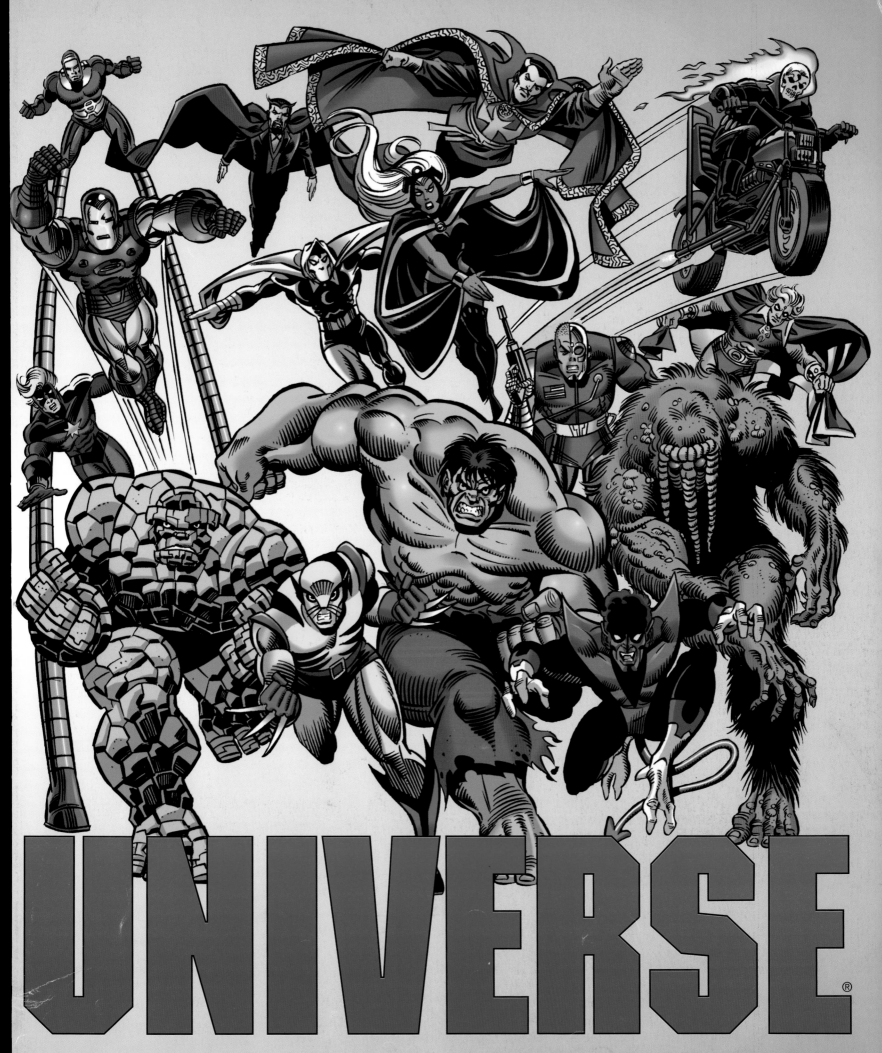

UNIVERSE

HARRY N. ABRAMS, INC., PUBLISHERS

The following illustrations were specially made for this book:

Page 2: *Super Heroes of the 1960s*. Pencils and inks: Mike Allred / Colors: Laura Allred and Bongotone

Page 3: *Super Heroes of the 1970s*. Pencils: Sal Buscema / Inks: Joe Sinnott / Colors: Bongotone

Pages 4–5: *Super Heroes of the 1980s*. Pencils and inks and colors: Michael Golden with Bongotone

Pages 6–7: *Super Heroes of the 1990s*. Pencils and inks: Dave Johnson / Colors: Bongotone

Pages 12–13: *The Fantastic Four: Marvel's First Family*. Pencils: Paul Ryan / Inks: Mark Farmer / Colors: Bongotone

Pages 40–41: *The Antiheroes: Human Torch, Sub-Mariner, and Hulk*. Pencils and inks: Jae Lee / Colors: Dana Moreshead and Bongotone

Pages 68–69: *Your Friendly Neighborhood Spider-Man*. Pencils and inks: Gil Kane / Colors: Bongotone

Pages 96–97: *Avengers Assemble!* Pencils: Steve Epting / Inks and colors: Tom Palmer and Bongotone

Pages 136–37: *Strange Tales: Heroes of the Supernatural*. Pencils: Gene Colan / Colors: Bongotone

Pages 156–57: *Protectors of the Universe*. Pencils: Ron Lim / Inks: Jerry Austin / Colors: Bongotone

Pages 178–79: *Vigilantes and Lawmen*. Pencils and inks: Klaus Janson / Colors: Dana Moreshead and Bongotone

Pages 208–9: *Mutatis Mutandis: The X-Men*. Pencils and inks: Andy Kubert / Colors: Electric Crayon

Editor: Eric Himmel
Designer: Dana Sloan
Photography: Zindman/Fremont

Library of Congress Cataloging-in-Publication Data

Sanderson, Peter
    The Marvel universe / Peter Sanderson.
        p.   cm.
    Companion vol. to: Marvel / Les Daniels. 1991.
    ISBN 0–8109–4285–2
    1. Marvel comic groups. 2. Comic books, strips, etc.—United States—
History and criticism. I. Daniels, Les, 1943– Marvel. II. Title.
PN6725.S36 1996
741.5'0973—dc20   95–7151

Published in 1996 by Harry N. Abrams, Incorporated, New York
A Times Mirror Company

Printed and bound in Japan

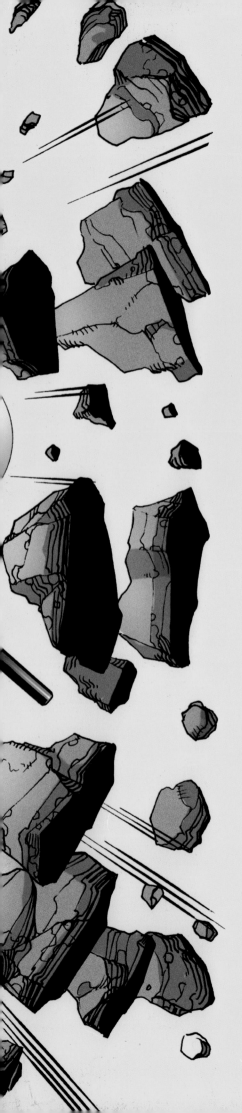

CONTENTS

FOREWORD: THE MARVEL UNIVERSE

The Marvel Universe is one of the most remarkable achievements of American popular culture in the late twentieth century: a single fictional cosmos with a history chronicled in thousands of comics. In many respects it is identical to our own world, but among the ordinary mortals of the Marvel Universe live the multitude of characters created by Marvel Comics, extraordinary figures who re-define heroic fantasy for an antiheroic age.

The comics super hero is essentially a familiar figure in a new guise, for throughout history humanity has been fascinated with stories about godlike and heroic figures; hence, the large bodies of Greek and Norse myths retain their dramatic power even though their religions are long dead. As has often been noted, the stories of the Marvel Universe constitute a modern-day mythology, equally vast in scope, whose heroes' strivings, usually represented through physical conflict against their foes, serve as metaphors for our own struggles in life on a grander scale.

Thanks to new mediums of storytelling like motion pictures, television, and comic strips, popular culture in this century is increasingly visual. The comic book, with its emphasis on color and simpli-fied forms, is particularly suited for the representation of human figures as powerful visual icons. Whereas in the past artists vied at depicting mythic figures like Venus and Apollo, now new generations of comics artists reinterpret the visual iconography of popular super heroes.

If the idea of the super hero strongly resonates in the modern American psyche, it is because comics writers have recast the traditional hero of adventure fantasy in terms appropriate to our time and place. The myths of the past depicted a world in which magic was real, but in the twentieth cen-tury people try to come to grips with the monumen-tal changes that the advances of science are making in their lives. Hence, science fiction has largely sup-planted magic as the mode of the fantastic. In super hero stories scientists become the creators of mirac-ulous wonders; radiation, not magic, either bestows phenomenal abilities upon a person or turns him into a monster.

One aspect of super hero stories that has partic-ular contemporary relevance is the theme of iden-tity: usually, comics super heroes have two. The impulse to change identity can be traced back to the many cultures in which a shaman or cult member would don a mask or costume to assume the aspect and powers of a god. The super hero's mask and costume serve a similar purpose, but Marvel stories also use the hero's double identity to symbolize the modern individual's divided self. Time and again in this book you will see characters whose dual per-sonae represent a sharp division between two unrec-onciled sides of their personalities.

The first comic-book super hero was DC's Superman, who debuted in 1938. A year later Marvel (or, as it was then called, Timely Publications) introduced its own super heroes. Even during this "Golden Age of Comics," which lasted until the first wave of popular fascination with super heroes had run its course in the late 1940s, there were "crossovers" between super hero series. In fact, Marvel was responsible for the first one, when the original Human Torch first battled the Sub-Mariner in *Human Torch* #5 (1941). Soon DC's heroes were regularly teaming up as the Justice Society of America, and after World War II Marvel's leading super heroes formed the All-Winners Squad. Still, outside these team books, it was rare for characters from one super hero series to appear in another: for the most part the heroes lived not in a common "universe," but in their own separate worlds.

In the late 1950s DC editor Julius Schwartz launched the "Silver Age of Comics" with his successful revivals of various Golden Age DC super heroes. In 1961 Marvel writer-editor Stan Lee, in creative partnership with such master comics artists as Jack Kirby and Steve Ditko, responded by creating a new "Marvel Comics" line of super heroes, beginning with the Fantastic Four. Lee's principal modus operandi was to ask himself what it would be like if his characters actually existed in reality. Would their personal lives run as smoothly as their victories over criminals? What would actually motivate them to pursue such dangerous careers? How would the world react to these masked beings of great power who had materialized in their midst? By taking this approach, Lee and his collaborators transformed super heroes from cardboard figures into vividly multidimensional characters whose triumphs and failures resonated with the experience of their readers.

Up until then, the status quo of a super hero series—and the personalities of its characters—rarely changed. Now they did. Lee and the writers who followed him could devise new stories based on those of the past, showing how events shaped the characters' lives and personalities. After thirty-five years Marvel characters have developed rich histories, as writers embroider what had gone before according to the changing tastes of new generations of readers, taking care every so often to recapitulate key events of the past. At times, writers would have to undo past story developments that had come to be recognized as mistakes, even finding imaginative ways to bring the dead back to life. But they always had to work within the rules: they could not just declare a story noncanonical, but had to find a way around it. The unspoken assumption everyone shared was that there was a classic tradition of Marvel stories that began with *Fantastic Four* #1 and that each story done since was part of the overall tapestry that grew larger and grander with each passing year. (At the same time, Marvel's Golden Age heroes were resurrected and carefully integrated into the mix of new characters.)

In part the Marvel Universe came about because it was convenient and suited Lee's style of working in the early 1960s, when he scripted virtually all the books. Almost all the characters were based in New York, so why couldn't they run into each other? Besides, to have the Fantastic Four guest star in *Amazing Spider-Man* #1 would presumably induce *Fantastic Four* readers into buying the new series.

MARVELS

Over the decades super hero comic books have so proliferated that regular readers may take the sight of superhuman beings for granted. Writer Kurt Busiek and artist Alex Ross's 1994 four-issue series *Marvels* triumphantly restored the sense of wonder to the genre. *Marvels'* strategy is not to center the story on the super hero's point of view but to see the heroes and villains—the "Marvels"—through the eyes of normal, everyday people and to show how the Marvels' existence affected their lives.

In particular *Marvels* follows the life of Phil Sheldon, a young photographer who was present at the first appearance of the first Marvel hero, the original Human Torch, in 1939. While photographing a battle between the Torch and the Sub-Mariner, Sheldon realized he was watching something that transcended the ordinary, something miraculous, and by witnessing and recording it, he had become part of the miracle. From then on Sheldon devoted his career to chronicling the history of the Marvels through his photography.

Through his painted artwork Alex Ross succeeds more effectively than any comics artist before him in combining the fantastic figures of the Marvels with an everyday Manhattan setting into one seamless, utterly credible whole.

In vividly recreating landmark events from stories Marvel published from 1939 to 1973, *Marvels* stands as a tribute to the enduring power of the classic stories and concepts on which the Marvel Universe is based.

Marvels #4 (1994) Script: Kurt Busiek / Art: Alex Ross

Lee and his collaborators clearly enjoyed the story possibilities of having their myriad characters meet since they had endowed all these characters with very different personalities.

It was not long before the term "Marvel Universe" could be taken literally. Whereas early super hero stories usually restricted themselves to the setting of a single fictional city, like Batman's Gotham City, Marvel stories ranged as far through space and time as the imagination would allow. The main super heroes were based in New York City, just as Marvel Comics is. But the stories range over the world, from real countries like Captain Britain's United Kingdom and the Red Skull's Germany to legendary places like the Sub-Mariner's Atlantis and new fictional creations like the Black Panther's Wakanda or Doctor Doom's Latveria. And from Earth the Marvel Universe extends throughout space, even including vast alien empires in other galaxies.

The scope of Marvel stories likewise reaches far into the past, with characters like the Eternals, who were mistaken for gods in ancient times; one of them, the Forgotten One, was Gilgamesh, hero of the ancient Babylonian epic. There were Marvel heroes in King Arthur's court and in the nineteenth-century American West. Actual gods and heroes from the myths of ancient cultures have been incorporated into the present-day Marvel Universe: the Norse thunder god Thor is the hero of his own series, while Hercules is a member of the Avengers.

Properly speaking, the Marvel Universe is a "multiverse" consisting of many universes, each existing in its own dimension. There are other-dimensional worlds inhabited by "gods," like Thor's Asgard, and surreal mystical realms like the Dark Dimension ruled by Doctor Strange's enemy Dormammu. Then, too, there are parallel Earths that had diverged from the principal Marvel Earth's reality sometime in the past. On these alternate Earths history can take a very different route, like the Earth of the Squadron Supreme, which pro-duced an entirely different set of super heroes who seized political control of the United States. There are even "pocket universes," alternate realities sharply limited in space or time.

Finally, Marvel stories range far into the future, which in the Marvel Universe has no one pre-destined route. Instead there are innumerable alter-nate paths that the future may take: one leads to the early career of humankind's potential savior, Cable, while another produces the dictatorial reign of Kang the Conqueror. Marvel even devised a whole family of books set in the year 2099 A.D. of an alternate future. The Marvel 2099 line was heavily influenced by cyberpunk fiction, with heroes who literally travel through cyberspace and an America dominated by an oligarchy of corporations. Within this world new figures assumed the roles of Marvel's archetypal heroes: thus, there arose a Spider-Man 2099 and even the X-Men of 2099.

Marvel series based on licensed properties—most notably *Conan the Barbarian*—do not fall within the purview of this book, nor do the creator-owned series published by Marvel's Epic line and stories set in other fictional universes such as those published by Marvel's sister company, Malibu. Moreover, literally thousands of characters have been created in the Marvel Universe over the last three and a half decades, to appear in even more thousands of stories. We don't have the space in this book to cover all of them, or to include every Marvel series or samples of work by every Marvel writer and artist. Instead, we seek to introduce the most significant characters in the Marvel Universe and describe the stories that have served as land-marks in its history.

Just as Marvel's innovations in characterization and storytelling have influenced every other work in super hero comics, so too has the concept of the Marvel Universe. Other comics companies have constructed a single fictional reality around their characters, but these efforts never work as well as the Marvel Universe, for two primary reasons. First, the Marvel Universe grew organically, with Stan Lee and his small band of artists working together, seamlessly integrating their new creations into their fictional world. Second, a consistent, unified fiction-al reality means nothing if no one cares about the characters who dwell within it. The greatest reason for Marvel's success was the brilliance of its charac-terizations. It is through Marvel's greatest characters that this book will explore the Marvel Universe.

THE FANTASTIC FOUR: MARVEL'S FIRST FAMILY

Despite the colossal success of super hero comics in the late 1930s and the 1940s, Superman, Batman, and Wonder Woman—all characters owned by National Periodical Publications, now known as DC Comics—were the only survivors of the collapse of the entire genre by 1951. Starting in 1956, however, National had staged successful revivals of a number of "Golden Age" heroes, beginning with the Flash. So it was that in 1961 Stan Lee's publisher, Martin Goodman, directed him to create a super hero team book that would outdo National's recent success, *Justice League of America*.

Goodman got something quite different. In part this was inevitable. The Justice League was a team of National's biggest super hero stars; the company that was to become Marvel did not have any star characters yet. An attempt to revive three "name" characters—Captain America, the Human Torch, and the Sub-Mariner— had failed in the 1950s, and Lee and his artists, notably Jack Kirby and Steve Ditko, were now busy turning out stories pitting sharp-witted humans against monsters from beyond, like Googam, Son of Goom, the Blip, and the talking dragon Fin Fang Foom. These are enjoyable stories with unintended camp appeal for adults, but they hardly tapped the talents of their creators. Stan Lee and Jack Kirby, whether they knew it or not, were about to make a sudden creative leap from these trivial fillers to the best work of their lives: stories that would reinvent the comic book adventure genre. Early on in its run, Stan Lee placed the label "The World's Greatest Comic Magazine" on the cover of *The Fantastic Four* as a bit of intentional hyperbole to grab the attention of new readers. Had he called it "The World's Most Innovative Super Hero Comic," he would not have been overstating the situation in the least.

THE FANTASTIC TWO: LEE AND KIRBY

Stan Lee had been in the comics business for two decades, starting with a prose story for the original *Captain America* series, and had grown tired of turning out material for juvenile audiences. As he has recounted the story, his wife challenged him to do the kind of comic book he himself wanted to read. Goodman's demand for a new super hero book provided him with the opportunity.

Jack Kirby was several years older than Lee and had worked in comics even longer. Kirby and his former partner, Joe Simon, had already jointly created Captain America as well as Manhunter, Fighting American, and an array of other super heroes for various companies; they made ventures into other genres after the 1940s super hero boom ended, such as *Boys' Ranch* and *Black Magic*, and together they invented the romance comics genre. In retrospect, it seems Kirby's entire career up until then was a preparation for his great work of the 1960s.

Hence *The Fantastic Four* was the result of a collaboration of two men who had spent decades mastering the conventions of story and art in the comic book medium, had grown dissatisfied with the limitations supposedly imposed upon them by the juvenile market, and were now, in their early forties, ready and willing to experiment, to re-create the super hero adventure story both for themselves and for a new generation of readers.

Lee and Kirby's extraordinary creative strengths became evident immediately. With rare exceptions, such as Will Eisner's *The Spirit*, comic book dialogue up until 1961 was never noteworthy for wit, for drama, for characterization, or for anything beyond simply getting the story told. Except for comic relief characters, there were no individual voices in the dialogue; a Justice League meeting read like one person holding a conversation with himself.

Lee often claimed Shakespeare as an influence, and he certainly had a playwright's instinct for using language both to move a story along and to establish character. Stan could shift in the same story among the lower middle-class New York dialect of the Thing, the scholarly tones of Mister Fantastic, the oratorical tirades of Doctor Doom, and the highly stylized speeches made by noble characters such as the Silver Surfer and godlike entities like Galactus. Furthermore, though Lee often verged into melodramatic excess, the emotional truth of his dialogue always came through.

Lee's dramatic scope in language was perfectly complemented by Jack Kirby's exceptional range in visualization. Kirby's first few years of *The Fantastic Four*, with pages divided into up to nine confined panels, only hinted at what was to come. Once past issue #30 Kirby began to let loose his firepower. Initially they were thin and unprepossessing; now the Fantastic Four became more muscular, more dynamic in movement, even sexier. Panels became fewer in number but greater in size, and battle scenes radiated power and speed that had never been seen in comics before. Whereas most comic book stories had often downplayed the fight sequences

NEVER *AGAIN* WILL YOU BE ABLE TO STRIKE YOUR *MASTER!*

ONE MERE BOLT OF *ANTI-MOLECULAR FORCE* WILL DISINTEGRATE YOU FROM THE FACE OF THE EARTH!

ONLY IF IT *HITS* ME, MADMAN!

--AND THAT'S NOT ABOUT TO *HAPPEN* WHILE I CAN STILL INSTANTLY *CHANGE* TO ANY FORM I CHOOSE!

in comparison to the overall plot. Marvel's battles had a visceral impact, visually and dramatically, that energized the entire story. More importantly, in collaborating on the plots, Lee and Kirby devised brilliant concepts and characterizations that continue to seize readers' imaginations to this day.

THE ORIGIN

Up until 1961 the conceptual basis for creating most super heroes was their special powers or some other gimmick: the Flash was the Fastest Man Alive, for example, and Green Arrow shot trick arrows with boxing gloves or other whatnot attached to the ends. But all of them had basically the same noble, do-gooder personality.

Lee's guiding principle in creating the first wave of Marvel heroes was to create characters with distinct personalities and ground the fantasy elements in a recognizable, believable reality. His characters would encounter the same problems that would exist in the real world, and they would react the way that real people would, the only difference being that they would possess super powers. This strategy gives the fantasy elements more dramatic power by giving readers realistic characterizations and situations with which they can empathize. Moreover, Lee and Kirby made it possible for later writers to build on their innovations, thus paving the way for the "mature" fantasy comics of the present day. It was Lee and Kirby who began the deconstruction of the super hero adventure in the 1960s,

and in the process they made that dying myth of popular culture live again.

So it is that when we first see the future Fantastic Four in the flashback to their origins in issue #1, they are bitterly quarreling with each other. Reed Richards, the leader of the team, has built a prototype spaceship that will enable them to reach "the stars" before their Cold War adversaries, the Soviets, can. Now the government has decided to shut down the project. Richards intends to make a test flight anyway to prove the project's worth. Ben Grimm objects that the ship isn't sufficiently shielded against radiation, but Sue Storm counters that Ben is acting like a coward. Grimm backs down, and that night the foursome, including Sue's younger brother Johnny, sneaks past security and takes off in the starship.

A pattern is established: Richards's team ultimately operated as a unit, but there was always the danger of dissension and disunity, with Ben as the particular source of turmoil. Indeed, it would come out that Ben was also a rejected rival with Reed for Sue's affections. Moreover, Richards and his team were not only fighting among themselves, but were setting themselves at odds with their government: although the "theft" of the spaceship had no legal repercussions, it was a foreshadowing of the antagonism between Marvel's heroes and society that lay in the near future.

With Grimm as the pilot, the ship runs up against Murphy's Law in the form of an unexpectedly intense storm of "cosmic radiation." They crash-

Fantastic Four #58 (1967)
Script: Stan Lee / Pencils: Jack Kirby / Inks: Joe Sinnott

Reed Richards demonstrates his stretching abilities in combat with his archenemy Doctor Doom.

Fantastic Four #64 (1967) Script: Stan Lee / Pencils: Jack Kirby / Inks: Joe Sinnott

Just as Stan Lee gave characters grandeur through language, so too Jack Kirby gave the world of the *Fantastic Four* an epic sweep. Over the course of just a few years his depictions of aliens evolved from the dwarfish, reptilian Skrulls to the imposing Galactus, his renditions of futuristic design from the Skrulls' hackneyed flying saucers to Galactus's monumental world ship. Kirby could move from beautifully constructing an elaborate medieval palace for Doctor Doom to inventing a unique alien architecture for the Great Refuge of the Inhumans to blending the supposedly "primitive" with the futuristic in the Black Panther's African kingdom. He would draw an intricately detailed piece of machinery merely to give the Thing something to lift as a throwaway gag (above left), or put together a unique collage—something his many followers have yet to imitate—to evoke the literally unearthly environments of alien worlds and dimensions (left). As master of so many comics genres, Kirby could depict comedy and romance as skillfully as he could the spirit of adventure. Although renowned for working on a larger-than-life scale, he could also evoke emotion through the subtlest of facial expressions and body language (below). He remains unmatched among comics artists in evoking a sense of wonder in his work.

Fantastic Four #49 (1966) Script: Stan Lee / Pencils: Jack Kirby / Inks: Joe Sinnott

Fantastic Four #51 (1966) Script: Stan Lee / Pencils: Jack Kirby / Inks: Joe Sinnott

Fantastic Four #1 (1961) Script: Stan Lee / Pencils: Jack Kirby / Inks: Unknown

land back on Earth only to discover that, in the tradition of 1950s science-fiction movies, they have been radically transformed by the radiation. Richards can now stretch his body as if it were a rubber band, Sue can turn invisible, and Johnny can burst into flame without any harm to himself. But all of them can resume normal human form at will; not so Ben, who mutates before them into a lumpish, orange-hued monster of enormous strength. Their shared experience binds them together, and in a burst of 1960s idealism, they decide—Ben, reluctantly—to use their new abilities for the good of humanity.

Although the first issue found the team based in Central City, as fictional a place as Superman's Metropolis, by issue #2 they had relocated to the very real New York City. They didn't have secret identities; being super heroes was the Fantastic Four's profession. They had incorporated themselves, had their own office complex in the midtown Manhattan Baxter Building (later rebuilt as Four Freedoms Plaza), and had even suffered financial and legal crises from time to time.

The absence of secret identities meant that the Fantastic Four did not split their psyches as other costumed heroes did between a repressed everyday identity and an aggressive masked persona: aside from the Thing, the FF had integrated, comparatively healthy personalities. The Fantastic Four did not even have costumes at first, and finally adopted utilitarian jumpsuits. Furthermore, Lee and Kirby dealt not only with their heroes' attempts to function in a

FANTASTIC FOUR ORIGIN

It seemed perfectly natural to super hero comics readers of the 1960s that as soon as characters acquired unusual powers, they would concoct colorful new names for themselves, as the FF did at the end of their first issue (above). Reed Richards, the leader of the team, rather grandly named himself Mister Fantastic. His ability to stretch his body as if it were made of rubber proved to be less important than his Einstein-level genius and his inexhaustible capacity to invent technology decades ahead of whatever this year's state of the art might be. Susan Storm, now Reed's wife, who called herself the Invisible Girl (and later the Invisible Woman) can turn herself and anything else invisible or surround herself with an equally invisible solid energy field. Her brother, Johnny Storm, a teenager at his debut and now, thirty years later, a young man in his twenties (normal aging by the conventions of comic book time), chose the name the Human Torch, becoming the successor to Marvel's first super hero of the 1930s; like his android predecessor, Johnny can burst into flame without harm to himself, and even fly. Finally, Ben Grimm, former pilot and astronaut transformed into a monster with enormous strength and a nearly impenetrable, rocklike hide, elected the name his former love Susan first blurted out upon seeing him: the Thing.

RICHARDS'S INVENTIONS

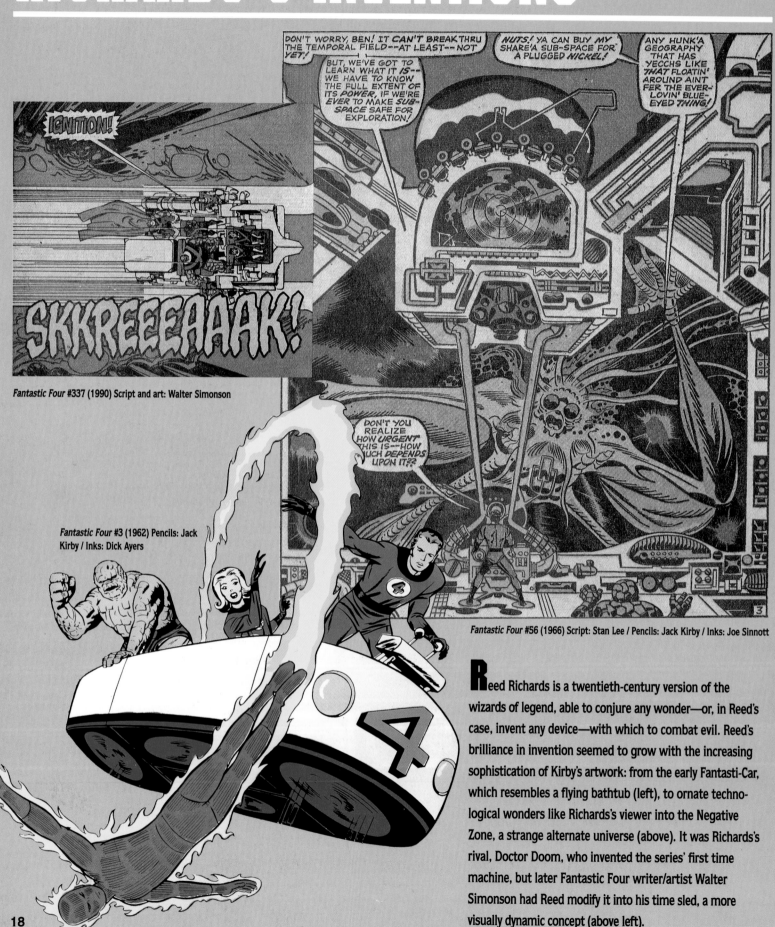

Fantastic Four #337 (1990) Script and art: Walter Simonson

Fantastic Four #3 (1962) Pencils: Jack Kirby / Inks: Dick Ayers

Fantastic Four #56 (1966) Script: Stan Lee / Pencils: Jack Kirby / Inks: Joe Sinnott

Reed Richards is a twentieth-century version of the wizards of legend, able to conjure any wonder—or, in Reed's case, invent any device—with which to combat evil. Reed's brilliance in invention seemed to grow with the increasing sophistication of Kirby's artwork: from the early Fantasti-Car, which resembles a flying bathtub (left), to ornate technological wonders like Richards's viewer into the Negative Zone, a strange alternate universe (above). It was Richards's rival, Doctor Doom, who invented the series' first time machine, but later Fantastic Four writer/artist Walter Simonson had Reed modify it into his time sled, a more visually dynamic concept (above left).

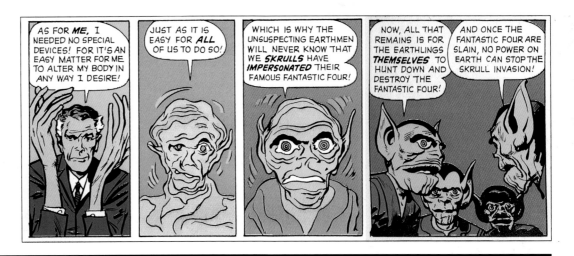

recognizable contemporary urban world, but also with that world's reaction to the strange super-humans in their midst. By their second issue the Fantastic Four had become media celebrities. Seven issues later the foursome even made a movie about their brand-new careers.

Lee and Kirby were well aware of the initial conflicts between society and Timely's first heroes, the Sub-Mariner and the original Human Torch, and they reintroduced a theme that would become a foundation of the Marvel super hero story: that beneath the admiration humanity has for super-humans—who stand for any individual or group that is different from the majority—lies the potential for distrust, fear, envy, and bitter, irrational hatred. In issue #2 the alien Skrulls use their shape-shifting powers to impersonate the Fantastic Four and frame them for various crimes, and the U.S. military is sent to hunt the team down. The Fantastic Four quickly cleared their names, but Lee and his collaborators would soon focus on this theme with Spider-Man, the Hulk, and, to a far greater extent, the X-Men.

THE THING

The member of the Fantastic Four who was the most original in conception was Benjamin J. Grimm, the inhuman-looking Thing. Before Grimm, super heroes were invariably perfect physical specimens; their special powers were gifts that readers envied. But Ben's powers were his curse: the Thing's colossal strength came at the price of his grotesque, rocklike

form. Though Ben likes to refer to himself as the "idol o' millions," he is all too aware that some people are repelled and even frightened by his appearance.

Worse, the Thing's freakish form made manifest the monster within himself. Thematically it was real-ly no surprise that the bitter, resentful malcontent on Richards's flight team would transform into a raging brute, lashing out at his teammates upon their return from space. (In his writing stint on the Thing's own series in the 1980s, John Byrne would reveal that Ben had been embittered as a child by the death of his brother in gang warfare on Manhattan's Lower East Side, and had ended up replacing him as gang leader; only the persistence of his uncle Jake reclaimed Ben from a possible life of crime.) His hair-trigger temper, his understand-able tendency to self-pity, and his very body all sep-arated him from any chance he had at happiness and made him hate himself and the world that had condemned him to such a fate. Like the Hulk two years later in the Avengers or Wolverine in the X-Men in the late 1970s, the Thing was the Fantastic Four's potential time bomb, a force for chaos held in check by the strictures of the team. Twice during Lee and Kirby's run on the series the nightmare came true: Ben was brainwashed by the Fantastic Four's foes into turning his rage and strength against his teammates.

Lee and Kirby began changing their initial char-acterization of the Thing with issue #8, in which he met the sculptress Alicia Masters. As a blind woman and an artist, she "saw" only the kind man beneath

Fantastic Four #2 (1962)
Script: Stan Lee / Pencils: Jack Kirby / Inks: Unknown

The alien Skrulls are part of a long line of fictional shape-shifting menaces going back to mythological characters like the demi-god Proteus. Able to impersonate anyone so as to subvert society from within, the Skrulls reflect the 1950s and 1960s cul-tural paranoia present in other science-fiction works like the 1956 film *Invasion of the Body Snatchers* and television's *The Invaders* (1967–68).

FANTASTIC FOUR VILLAINS

THE MOLE MAN, from *Fantastic Four* #88 (1969) Script: Stan Lee / Pencils: Jack Kirby / Inks: Joe Sinnott

BLASTAAR and **ANNIHILUS**, from *Fantastic Four* #289 (1986) Script and pencils: John Byrne / Inks: Al Gordon

PSYCHO-MAN, from *Fantastic Four* #282 (1985) Script and pencils: John Byrne / Inks: Jerry Ordway

Lee and Kirby devised a wide and varied assemblage of colorful adversaries for Marvel's premier super hero team, beginning with the Mole Man, their opponent in *Fantastic Four* #1. Alicia's stepfather, the Puppet Master, controls his victims through mystical puppets, while Diablo relies on ancient alchemy. Blastaar and Annihilus are menaces from the Negative Zone, and Psycho Man, who manipulates emotions, inhabits a "microverse." The shape-shifting alien known as the Impossible Man was originally a nuisance with his continual pranks, but eventually he ended up starring in his own comedy series.

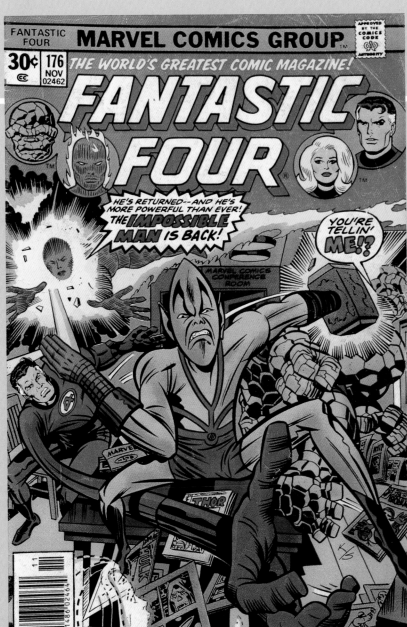

THE IMPOSSIBLE MAN, from *Fantastic Four* #176 (1976) Pencils: Jack Kirby / Inks: Joe Sinnott

DIABLO, from *Fantastic Four* #232 (1981) Script and art: John Byrne

THE PUPPET MASTER, from *Fantastic Four* #8 (1962) Pencils: Jack Kirby / Inks: Dick Ayers

Fantastic Four #1 (1961) Script: Stan Lee / Pencils: Jack Kirby / Inks: Unknown

Fantastic Four #68 (1967) Script: Stan Lee / Pencils: Jack Kirby / Inks: Joe Sinnott

BEAUTY AND THE BEAST

In early issues and sometimes even today Ben tries to conceal his hideousness beneath scarves, masks, hats, and even helmets, but ultimately to no avail. Especially early on in the series, people would be horrified by the very sight of him. Ben's own revulsion at his physical appearance may be greater than anyone else's, founded in his fear that he is less than a fully human being and unworthy of being loved by Alicia or anyone else. But Ben's capacity for rage and self-loathing is the other side of his equally strong capacity for love and sensitivity.

Fantastic Four #274 (1985) Art: John Byrne
The Thing gives voice to his favorite catchphrase.

the rough exterior. In having Ben fall in love with Alicia, Lee and Kirby devised a variation of the centuries-old myth of Beauty and the Beast, but with a new twist. Even though Ben longed to be human so he could marry her, he could not help wondering, as Lee and later Byrne noted, whether Alicia loved his "real," ordinary human self or merely his powerful facade.

Eventually, Ben's bitterness toward Reed turned into good-natured and useful verbal jabs at his old friend's unconscious pomposity. Ben's continual fights with the Human Torch are now usually played for laughs, as if they were two children squabbling. His real fits of anger are directed against worthy targets, such as Doctor Doom, but when they were

not, they were depicted as childish temper tantrums for which Reed or Sue scolded him. Ben's jokes deflate the melodramatic pretensions of characters around him, and he keeps spirits high even amid the greatest suspense. He is comedy itself charging into battle with his invariable war cry, "It's clobberin' time!"

THE FAMILY AND THE PSYCHE

The kind of team the Fantastic Four became reflects the mature perspective of its creators; Lee and Kirby were not here designing yet another self-centered power fantasy about the young man with a heroic career he hides from his loved ones, or even a book about a "club" of heroes like DC's Legion of Super Heroes. In fact, the FF was a surrogate nuclear family, with Reed and Sue as the parents and lovers. Johnny was literally Sue's brother and Reed's brother-in-law, but figuratively he was also their son,

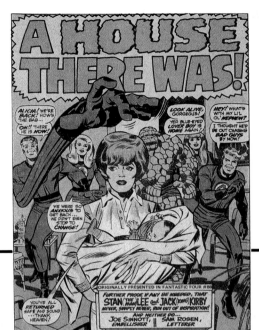

Fantastic Four #88 (1969) Script: Stan Lee / Pencils: Jack Kirby / Inks: Joe Sinnott

The familial nature of the Fantastic Four was emphasized by the birth of Reed and Sue's son Franklin, named after Sue and Johnny's late father.

maturing over the years from a rebellious juvenile to a young adult. (He ended up marrying an alien Skrull named Lyja the Lazerfist who used her shape-changing powers to resemble Alicia Masters.) Ben was Reed's lifelong friend and partner and hence a surrogate brother. When Reed and Sue once took a leave of absence from the team, it was Ben who assumed the role of leader and father figure. Yet Ben also has a mock sibling rivalry with Johnny, and in his more juvenile moments, Ben can seem no more than a cranky child throwing a temper tantrum. At such times he too becomes a "son" in the Fantastic Four's family dynamic.

In comic books from the 1930s into the mid-1960s the status quo of a series was inviolable: Clark Kent would never marry Lois Lane. Now, in the 1990s, Kent and Lane are engaged, and they have the Fantastic Four to thank. The very first super hero wedding was that of Reed Richards and Sue Storm in *Fantastic Four Annual* #3, and their son Franklin was born three years later in *Annual* #6.

Comics writers now generally acknowledge that the Fantastic Four is a family, but the foursome also represents different aspects of a single mind. Reed is intellect; the other three, in their different ways, represent emotionality. Reed's wisdom and maturity, Ben's spontaneity and earthiness, Johnny's youth and rebelliousness, and Sue's love and compassion together constitute an image of a whole, fully rounded human personality.

This is why although there have been many sub-

Fantastic Four #259 (1983) Script and art: John Byrne

Lee and Kirby soon discovered that Ben had great potential as a comedian. It is no surprise, really, that the Thing has only four digits on each hand like Bugs Bunny or Mickey Mouse. He has a cartoonlike appearance with a face that easily conveys exaggerated expressions; from one point of view the Thing is grotesquely ugly, but at the same time there is something appealingly humorous about him. Thus, here, during an otherwise deadly serious battle, Ben gets knocked through a supermarket in a panel designed as a pop art sound effect (an example of artist John Byrne's frequent experimentation with his medium).

Fantastic Four #51 (1966) Script: Stan Lee / Pencils: Jack Kirby / Inks: Joe Sinnott

"THIS MAN . . . THIS MONSTER!"

The Thing's finest hour occurred in the story "This Man . . . This Monster!" in issue #51, a story that illuminates Ben Grimm's character even though he is virtually a secondary player throughout. The tale opens with the Thing alone in the city, consumed by despair over his fate to live his life as a monster, wishing desperately to be freed from his curse. He gets his wish when he falls victim to an unnamed scientist who envies the great power that Grimm wants to give up and who turns himself into the Thing, returning Grimm to human form. The impostor usurps Ben's place in the Fantastic Four, planning to take the team over from Richards, a man he hates: it is as if, in the guise of this impostor, Ben's dark side has come to life, freed of his human conscience. But then Reed, blinded by his single-minded pursuit of knowledge, becomes entrapped in the Negative Zone, the strange dimension he has just discovered. Moved by witnessing Richards's own selflessness and courage, the impostor sacrifices his life by entering the Negative Zone to save Richards. Back on Earth the real Ben reverts to his monstrous form, and Reed, realizing what has happened, speculates that in taking on the Thing's form, the impostor had also discovered within himself Ben's great courage. This is a tale about the capacity for heroism within every man, the potential for redemption even for the direst sinner, and is a moving statement of the theme of the link between power and responsibility that lies at the heart of Marvel comics.

Fantastic Four #51 (1966)
Pencils: Jack Kirby / Inks: Joe Sinnott

stitute members of the Fantastic Four over the years, ranging from the Inhuman Crystal to the She-Hulk, none of the additions ever prove to be permanent. The personality mix in the original Fantastic Four is perfectly balanced and complementary as it stands.

As we shall see with the other great Marvel heroes of the 1960s, perhaps the most important aspect of the Marvel revolution was that lead characters were based, consciously or not, on psychological archetypes. This is why as tastes change and generations pass, the essential appeal and dramatic strength of the classic Marvel characters remain vital and contemporary.

DOCTOR DOOM

Although the series is titled *The Fantastic Four*, there is a fifth character who is nonetheless essential to its success: the archvillain Doctor Victor Von Doom, introduced in issue #5. While Reed, Sue, and Johnny embody youth, idealism, and vitality, Doom is a morbid visual icon of evil, hiding a scarred visage behind a death's-head mask. His massive suit of armor makes him look more like a robot than a human being, anticipating in appearance the cyborgs of later popular culture, especially *Star Wars'* Darth Vader.

Whereas the Fantastic Four are based atop a skyscraper in the nation's largest city and represent modern America, Doom speaks for Europe and an almost medieval past. Clad in armor, enthroned in a storybook castle, he rules the anachronistic Eastern European pocket nation of Latveria as its absolute monarch. The villagers are submissively grateful for the security he has brought them, even as his ominous robot guards stalk their villages. It is Doom's goal to extend his repressive paternalistic rule to the entire world and to return contemporary America to the nightmare of European tyranny its residents had fled.

Yet, strangely, Doom was also a victim of that very tyranny. The son of Gypsies who died under persecution, Doom vowed revenge and threw himself into the study of both his mother's sorcery and the science he learned in American schools. But, like the child of abusive parents who grows up to abuse his own children, Doom returned to Latveria and seized the throne, only to become a tyrant himself.

If Reed Richards was the modern scientist as hero, embodying the idealism of the space race of the early 1960s, Doom was the dark face of science, the side that created horrible weapons of destruction without considering the consequences.

Doom represented the road that Reed did not take. They were once both students at the same university, and Doom had nearly become Reed's

Fantastic Four #66 (1967) Script: Stan Lee / Pencils: Jack Kirby / Inks: Joe Sinnott

Fantastic Four #257 (1983) Script and art: John Byrne

FOUR INTO ONE

The various members of the Fantastic Four can be compared to different aspects of the human mind. Reed, when laying down the law to the others, is the superego; hot-tempered Ben and impetuous Johnny are the id. Sue is Reed's anima, without which he is incomplete. She openly expresses her love with ease, whereas he tends to keep his passions under tight control. Ben is Reed's shadow self, representing the potential for humor, anger, and violence that he denies himself. The Thing can vent a savage fury that Reed would ordinarily never allow himself (top), but can also express a joyful exuberance that perhaps Reed will never feel as strongly (above).

DOCTOR DOOM

Fantastic Four #87 (1969) Script: Stan Lee / Pencils: Jack Kirby / Inks: Joe Sinnott

In Doctor Doom Lee and Kirby combined numerous archetypal figures: the mad scientist, like Dr. Frankenstein, with the evil sorcerer of myth; the oppressive king with the modern totalitarian dictator; and the dark, armored knight with the brooding, unappreciated genius. He is a study in contrasts: his storybook castle (below) contains a high-tech laboratory (opposite), his outward elegance and refinement coexist with his ruthless lust for power; and, as with the Phantom of the Opera, his fearsome mask and costume conceal a man both physically and emotionally scarred. The full-page shot of Doom (left) is from a scene in which Sue and Crystal, rushing through his castle, suddenly come face-to-face with him across a long dinner table; it anticipates a similar sequence featuring Doom look-alike Darth Vader in *The Empire Strikes Back*.

MEANWHILE, ON THE OTHER SIDE OF THE WORLD, THE ABSOLUTE MONARCH OF *LATVERIA* GIVES HIS VISITOR FROM SPACE A RARELY-REVEALED GLIMPSE OF THE MOST AMAZING EXPERIMENTAL WORKSHOP-COMPLEX ON THE FACE OF THE GLOBE--

FEW LIVING BEINGS HAVE EVER BEEN PRIVILEGED TO WITNESS THIS SIGHT!

IT IS ONE OF THE MOST *CLOSELY-GUARDED* INSTALLATIONS IN EXISTENCE!

FOR, IT IS *HERE* THAT SOME OF MY GREATEST SCIENTIFIC PROJECTS ARE CARRIED OUT!

THOUGH MUCH OF THIS IS *STRANGE* TO ME, I SENSE THAT YOUR *PURPOSE* IS ONE OF WORLD-WIDE *DESTRUCTION!*

MY EYES DO NOT VIEW THIS SCENE WITH *FAVOR!*

Fantastic Four #57 (1966) Script: Stan Lee / Pencils: Jack Kirby / Inks: Joe Sinnott

Fantastic Four #258 (1983) Script and art: John Byrne

dormitory roommate. But Doom, already locked within his own arrogance, rejected Richards's attempts at friendship, and Ben Grimm became Reed's roommate instead. Thus Ben, rather than Doom, became part of the surrogate family and group psyche that would become the Fantastic Four.

Reed's saving graces were his need for friends and a lover and his sometimes comical lack of ego. Ben and Sue are forever haranguing Reed to keep him from withdrawing from the world and losing himself utterly in his studies. Unlike Reed, Doom is driven by his colossal ego and his need to dominate the rest of humanity. Reed often seems unemotional, but Doom is constantly on the verge of rage, exploding in fury like a dangerous child at the least annoyance by one of his servants. But ultimately rage is all Doom has. In *Fantastic Four Annual* #6, Reed surprises Ben and Johnny with his rare, near-hysterical anger as he desperately strives to save the lives of the wife and unborn son he loves. However, like Shakespeare's Richard III, Doom renounced love, in his case that of his childhood friend Valeria, to devote himself to the empty pursuit of power and of vengeance against the entire world. Hence it is only right that Doctor Doom story lines often focused on the struggle for dominance between the two men, with Doom once even psychically taking over Richards's own body.

Before Doom the typical comic book super-villains were bank robbers; even many of the would-be world conquerors seemed somehow petty. What elevated Doom above the rest was that

O! For a Muse of fire, that would ascend
The brightest heaven of invention;
A kingdom for a stage, princes to act
And monarchs to behold the swelling scene.
— Shakespeare, *Henry V*

TO BE CONTINUED

Doom 2099 #1 (1993) Script: John Francis Moore / Art: Pat Broderick

DOOM 2099

In *Doom 2099*, created by John Francis Moore and Pat Broderick, a scientific genius who believes himself to be the original Doctor Doom (and may indeed be correct) has reappeared in the late twenty-first century in a youthful body and set about mastering the advanced science of his new age to impose his kind of order on a world torn by strife. Eventually, people who may or may not be the original Fantastic Four likewise turned up in a companion series, *Fantastic Four 2099*, originally written by Karl Kesel and drawn by Rick Leonardi.

he had made himself a king and there was indeed a sense of majesty about him; he was like a fallen angel who now ruled over his own hell. It was necessary for the Fantastic Four, representing American everymen, to thwart Doom's ambitions, but nevertheless one could not help feeling that the world would be a lesser place without the greatness of Doom in it.

EPIC ADVENTURES

Although the Fantastic Four contend with a wide variety of memorable adversaries, they are essentially not crimefighters but explorers. With the Fantastic Four as our guides Lee and Kirby took their readers to extraordinary places, filling the newborn Marvel Universe with a true sense of wonder. There was Subterranea, the underground realm of caverns that was the kingdom of the Mole Man, a nearly blind misfit from the surface world who rules legions of gnomelike humanoids called Moloids; the Sub-Mariner's undersea kingdom of Atlantis; the strange "subatomic" universe of Psycho-Man; and the dimension called the Negative Zone, a realm of death that nearly claimed Reed Richards more than once and that spawned the insectlike Annihilus, who sought to destroy all living beings in his paranoid fear they might kill him first.

In issue #13 the Fantastic Four journeyed to the moon, where they discovered the Blue Area, a zone with its own breathable atmosphere, mysterious ruins, and a great building with alien architecture.

Fantastic Four #13 (1963) Script: Stan Lee / Pencils: Jack Kirby / Inks: Steve Ditko

The Watcher makes his first appearance, interrupting a battle between the Thing and the super-powered primates commanded by his Soviet foe, the Red Ghost.

the home of Uatu, a member of the race of Watchers. His purpose in life was to observe the "inferior" beings of Earth but never to interfere with their activities. At first eerily inhuman in appearance and disdainful of humanity, in later stories the Watcher came to resemble a pudgy but harmless human giant as he developed an interest in the Fantastic Four's welfare. Despite his great power he was forbidden by the laws of his race from aiding the Fantastic Four directly, but nevertheless the Watcher repeatedly warned them of perils and guided them to the means of combating them.

Hence, the Watcher seems a metaphor for a benign but rather distant god, interested in humanity only as a scientist might be in the microscopic specimens he studies, and, like Wagner's Odin, bound by his own rules from intervening in their affairs even if he should so desire. The Fantastic Four see Uatu as a friend, but he is a being who helps only those who are capable of helping themselves. The Watchers' vow of noninterference came about when the young Uatu had persuaded his immortal race to impart the knowledge of nuclear energy upon a more primitive alien race, who thereupon used it to construct weapons with which they destroyed their own civilization in a planetwide war. They were henceforth paralyzed from acting to bring about good by their fear of bringing about evil. (Elsewhere in 1960s popular culture, the same theme crops up repeatedly in *Star Trek*.) Uatu bent these rules slightly when he bestowed his blessing upon the wedding of Reed and Sue Richards by enabling

Reed to find an alien device to dispose of the many party-crashing villains disrupting the occasion.

Uatu played his most decisive role in the Fantastic Four's lives in the Galactus trilogy (*FF* #48–50), the high point of the Lee-Kirby Fantastic Four collaboration. It begins as the kindly but mostly ineffectual Watcher warns the Fantastic Four of the approach of Galactus, the devourer of the "life energy" of planets, a wrathful god recast in science-fiction terms. To him human beings are irritating

Fantastic Four #48 (1966) Script: Stan Lee / Pencils: Jack Kirby / Inks: Joe Sinnott

Manhattanites fear the end of the world when the sky bursts into flame. Indeed, this biblical vision is a sign that Galactus is coming to destroy the planet.

Fantastic Four #48 (1966) Script: Stan Lee / Pencils: Jack Kirby / Inks: Joe Sinnott

Fantastic Four #50 (1966) Script: Stan Lee / Pencils: Jack Kirby / Inks: Joe Sinnott

The Silver Surfer, as drawn by Jack Kirby, was far from the 1960s Southern California cliché his name suggests. Rather, he was like an angel, recast in the visual terms of the super hero genre. Near the beginning of the Galactus trilogy (top panel), the Surfer maintains an Olympian silence in the face of human despair. By the end of the story (bottom panel), he has turned on his master and become humanity's protector.

ants or gnats of no spiritual worth; their homeworld is his prey. His arrival on Earth is marked by fire in the sky and other signs in the heavens that terrify the populace of Earth.

Just as angels blowing trumpets are to herald doomsday, so too Galactus had his own herald, the Silver Surfer. A lithe, powerful figure seemingly sculpted from pure silver, with an enigmatic, blank-eyed gaze and impassive features, the Surfer was at once an alien and an idealized human. Soaring through the air on his board, the Surfer inspired awe.

Seemingly emotionless, the Surfer shared his master's contempt for humanity and was his willing accomplice in its destruction. After being hurled to Earth in battle with the Fantastic Four, however, the Silver Surfer was found by the Thing's friend Alicia

Masters. Moved by her compassion for him and for the rest of mankind, the Surfer decided that humanity was worthy of survival and challenged Galactus in combat.

Ultimately, though, in Lee and Kirby's worldview, humankind could not rely on unearthly Redeemers but had to prove their own worthiness. Guided by the Watcher to Galactus's planet-size starship base, the Human Torch retrieved the world-devourer's Achilles' heel, a tiny weapon called the Ultimate Nullifier, capable of obliterating both Galactus and the universe with him. Wielding the Nullifier, Reed Richards, as Earth's representative, forced Galactus to acknowledge humanity's right to existence and to vow never to threaten the planet again. Unable to take vengeance on mankind for his defeat, Galactus instead directed his rage against his formerly "beloved" Surfer, exiling him to Earth. Little is permanent in super hero comics, and despite his pledge, Galactus would return time and again to menace Earth.

The Silver Surfer continued to evolve as a character. Lee and Kirby now characterized him not as an amoral accomplice of Galactus, but as a literally otherworldly innocent. If his first fall came after he learned from Alicia what it meant to be human in a

Fantastic Four #48 (1966) Script: Stan Lee /
Pencils: Jack Kirby / Inks: Joe Sinnott
Galactus's first appearance. It seemed odd that
an alien would wear a "G" on his chest, and this
was subsequently dropped from his costume.

Fantastic Four #262 (1984)
Script and art: John Byrne
John Byrne later established that
Galactus was not truly a humanoid
giant but was a force of nature,
perceived by members of each
sentient race in its own image.

Marvels #3 (1994) Script: Kurt Busiek / Art: Alex Ross

THE COMING OF GALACTUS

Marvels #3 (1994) Art: Alex Ross

As Kurt Busiek and Alex Ross make clear in their retelling of this story in the 1994 series *Marvels* (opposite and above), the coming of Galactus is nothing less than the arrival of the Day of Judgment, the End of the World. In their 1988 collaboration *The Silver Surfer: Parable* (right), Stan Lee and the great French cartoonist Moebius varied the theme further, depicting a religious movement that welcomed Galactus's new descent to Earth as if it were the Second Coming.

The Silver Surfer: Parable (1988) Script: Stan Lee / Art: Moebius

GALACTUS ORIGIN

"AND IN ALL THAT WORLD ONE MAN ALONE HAD THE WISDOM TO KNOW WHAT IT WAS THAT SO DOOMED THEM, AND THE WORDS TO SPEAK ITS NAME...

THE *PLAGUE* THAT THREATENS US ALL HAS NO CURE, FOR IT IS THE FINAL PLAGUE OF *ENTROPY!*

OUR WORLD, OUR VERY UNIVERSE IS GRINDING TO ITS *END!*

"HIS NAME WAS *GALEN,* AND THO' HE KNEW IT NOT, HE WOULD BECOME *GALACTUS!*

Lee and Kirby eventually revealed Galactus to have once been a mortal named Galen, the last survivor of a dying universe that preceded our own. Galen, shown above in a retelling by John Byrne, is transported by rocket to the center of the big bang that created the present cosmos (right). Somehow, he is transformed into a physical incarnation of destruction (bottom right), whose coming serves as a trial by fire for the population of inhabited worlds.

"THUS DID GALEN AND A SMALL CREW OF BRAVE MEN LEAVE THE WORLD OF TAA, SEEKING OUT THE CENTER OF THEIR UNIVERSE...

"...THAT PLACE WHERE *TIME* AND *SPACE* ENDED AS ALL THINGS RETURNED FROM WHENCE THEY HAD SO LONG SINCE COME.

"THE UNIVERSE BECAME A GIANT *BLACK HOLE,* CONSUMING ITSELF.

"AND INTO THAT UNIMAGINABLE CENTER PLUNGED THE *LAST OF TAA,* ALL THERE TO *PERISH...*

"...ALL SAVE *ONE.*

"AND TO HIM A VOICE WHISPERED, A VOICE BEYOND OUR COMPREHENSION: *THE SENTIENCE OF THE UNIVERSE!*

"NO BEING CAN GUESS HOW LONG THE MAN ONCE CALLED GALEN LAY IN THE EMBRACE OF A DYING COSMOS.

"BUT WHEN THAT DEATH GAVE ITSELF AS BIRTH TO *OUR* UNIVERSE, THE LAST SHIP OF TAA WAS FLUNG OUT--SOMEHOW REMADE!

"IT, AND ITS LONE SURVIVOR WAS EVENTUALLY FOUND BY ONE OF THE WATCHER'S RACE.

"...THEN CAME THAT FATEFUL DAY WHEN THE CUBE WAS *OPENED...*

"STUDIED, BUT DID NOTHING TO INTERFERE WITH ITS GROWTH, ALLOWING IT TO LEAP AGAIN TO SPACE, RESHAPING THE LAST SHIP INTO A GREAT *INCUBATOR!*

"LONG DID THAT SILENT CUBE DRIFT, AS THE UNIVERSE GREW AROUND IT, AND LIFE SPREAD.

"...AND GALACTUS LIVED!"

Fantastic Four #262 (1984) Script and art: John Byrne

Fantastic Four #55 (1966) Script:
Stan Lee / Pencils: Jack Kirby /
Inks: Joe Sinnott

The Silver Surfer pays for his rebellion
against Galactus by being exiled from
the heavens and trapped on the world
of the mortals he had helped to save.

Fantastic Four #57 (1966) Script: Stan Lee /
Pencils: Jack Kirby / Inks: Joe Sinnott

Doctor Doom achieved his greatest triumph
in usurping the power of the Silver Surfer,
thereby raising himself to the level of a god.

positive sense, he would fall again in learning
humanity's potential for evil. In issue #57, Doctor
Doom tricks the all-too-gullible Surfer and steals his
cosmic powers, reducing him to utter helplessness.
In the course of the story, the Surfer regains his
powers, but he has clearly reached a turning point,
and he demolishes his captor's palace in his wrath,
his eyes now opened to mankind's dark side. Soon,
the highly popular Silver Surfer moved into his own
series, written by Stan Lee and drawn by John
Buscema, which is chronicled in chapter seven.

KINGS OF OTHER REALMS: THE BLACK PANTHER AND THE INHUMANS

Lee and Kirby had introduced an African-American
character, Gabe Jones, as a member of Sgt. Fury's
Howling Commandos in their war comic series of
the same name in 1963. In 1966, in *Fantastic Four*
#52, they introduced the first black super hero, the
Black Panther (who predated—and has outlasted—
the 1960s radical political party of the same name).
The Panther became the Fantastic Four's staunch
ally and a fixture in the Marvel Universe to this day.

Strangely enough, the Panther's origin paral-
leled that of Doctor Doom. T'Challa, as the Panther
was known, was born into an isolated African tribe
in the fictional land of Wakanda, which had no contact with modern Western civilization. When he was
a boy his father, like Doom's, died at the hands of
an oppressor: T'Challa's father, the tribal chief, was
murdered by an American, Ulysses Klaw, who had

Black Panther: Panther's Prey #1 (1991) Art: Dwayne Turner

After Lee and Kirby the writer who did the most to develop the Black Panther was Don McGregor, whose "Panther's Rage" story line in T'Challa's own series in the 1970s portrayed a man of human vulnerabilities to pain and self-doubt, who nevertheless fought on in a lonely struggle against his former friend, the usurper Erik Killmonger. The Panther has frequently been used as an iconic warrior against racism, as in McGregor's 1980s story line in *Marvel Comics Presents* sending him to find his long-lost mother within a South Africa still in the grip of apartheid.

Fantastic Four #240 (1982) Script and art: John Byrne

Despite his normal appearance, the strangest of the Inhumans was perhaps their monarch, Black Bolt (standing in the center of the panel), a figure of nobility mixed with utter repression. He could never speak, not to scream in pain or to tell his beloved cousin Medusa how he felt for her, because of the incredible power of his voice: a mere whisper could shatter an entire city. Medusa, who finally became his wife, stands behind him, identifiable by her long hair, which she can mentally manipulate as a weapon. Behind her are other members of the Inhumans' Royal Family: the martial-arts master Karnak (with mustache); beside him the taller, bearded Gorgon, who can create earth tremors by stamping his hoofed feet; and the scaly, water-breathing Triton.

Fantastic Four #334 (1989) Script and art: Walter Simonson
Writer/artist Walter Simonson took the reins of the *Fantastic Four* series for a brief but memorable run in 1989 and 1990.

come to plunder Wakanda of the rare metal vibranium. Like Doom, the young T'Challa vowed vengeance, studied science abroad as a youth, and then returned to his homeland as its monarch. On his return from the West T'Challa underwent a ritual that earned him the right to wear the mask and costume of the Black Panther, the earthly representative of the patron god of the Wakandans; thus Lee and Kirby linked the contemporary motif of the super hero with the uses of masks and costumes in other older cultures.

Like Latveria, Wakanda became a strange mix of the old, with its retention of tribal ceremony and dress, and the futuristic, with T'Challa's high-tech palace. Unlike Doom, however, T'Challa was not consumed by his desires for revenge and domination. Rather, he used his genius for invention and the proceeds from selling vibranium to the outside world to transform Wakanda into a prosperous nation that surpassed other nations in its scientific achievements.

The Inhumans, introduced in *Fantastic Four* #45, were the second Lee and Kirby variation on a theme they began in 1963 with the X-Men: the idea of a hidden race of superhumans living among us, divided into warring factions of good and bad. Eventually it was revealed that the Inhumans were descended from the apelike species that would evolve into Homo sapiens. An alien race called the Kree, who would menace the Fantastic Four in contemporary times, performed genetic experiments with some of these ape-men, speeding up their evo-

lution. The Inhumans built a civilization that remains far in advance of "normal" mankind and discovered the mutagenic "terrigen mist" that they used to bestow superhuman powers, and often inhuman physical appearances, upon the members of their race.

Hence, while some Inhumans, like the young and beautiful Crystal, could easily pass as "normal" human beings, others, such as the water-breathing Triton, with his scales and fins, or Medusa, with her "living" hair, could not.

Again, Kirby and Lee's skill as storytellers lay in linking their fantasy creations to human realities that inspired reader identification. The key here was in starting a Romeo and Juliet–style romance between Crystal and Johnny Storm, which became imperiled

JOHN BYRNE

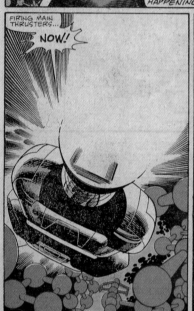

Fantastic Four #282 (1985) Script and art: John Byrne / Inks: Jerry Ordway

Fantastic Four #286 (1986) Script and pencils: John Byrne / Inks: Terry Austin

Fantastic Four #252 (1983) Script and art: John Byrne

Just as Kirby's work in the 1960s affected the generations of super hero artists that followed him, so now the new crop of artists in the 1990s invariably name John Byrne, along with Kirby and Frank Miller, as one of their primary influences. A great admirer of Kirby himself, Byrne combined within his own work masterful composition, a sharp sense of pacing, and an extraordinary gift for portraying characterization through facial expression and body movement. As writer and artist Byrne left his mark on many Marvel series, including *Fantastic Four*, *The Uncanny X-Men*, *The Incredible Hulk*, and *Namor*, as well as revamping DC's *Superman*. These pages from *Fantastic Four* show Byrne celebrating the theme of intrepid exploration of new worlds, from the subatomic cosmos of Psycho-Man (above) to deep space (above right). One of his unusual experiments was his "sideways issue" (right): one had to turn the comic ninety degrees to read it and was rewarded by panel layouts reminiscent of wide-screen movies.

Marvels #2 (1994)
Script: Kurt Busiek /
Art: Alex Ross

Marvels presented the wedding of Reed Richards and Sue Storm as the Marvel Universe's equivalent of a royal wedding, complete with the Beatles and other icons of 1960s

From top to bottom:
Fantastic Four #68 (1967)
Script: Stan Lee / Pencils: Jack Kirby / Inks: Joe Sinnott

Fantastic Four #384 (1994)
Script: Tom DeFalco / Pencils: Paul Ryan / Inks: Dan Bulanadi

Fantastic Four #387 (1994)
Script: Tom DeFalco / Pencils: Paul Ryan / Inks: Dan Bulanadi

Being married to Reed can make Sue seem staid herself, so sometimes artists try to shake up her image by giving her a new costume.

when it was revealed that she was the youngest member of the Royal Family of Inhumans, hiding in exile among normal humans, whose mores forbade contact between the two races.

The Royal Family took Crystal back to its futuristic Shangri-la, the Great Refuge of the Himalayas, where the Fantastic Four caught up with them and helped their monarch, Black Bolt, overthrow his usurper brother, the insane Maximus. Crystal would eventually return to the outside world and Johnny, and even join the Fantastic Four on two occasions as a temporary replacement for Sue (as did Medusa at another time). But neither her membership nor her romance with Johnny would last.

BEYOND THE SIXTIES

Jack Kirby left *The Fantastic Four* with issue #102 in 1970 to create *The New Gods* and its related titles at DC. Stan Lee remained on the series as writer for only another year; his collaborations with other artists on the book lacked the spark he had found with Kirby.

Numerous leading writers and artists have worked on what was "The World's Greatest Comic Magazine" over the more than twenty years since then. The most memorable runs of issues were the work of John Byrne and Walter Simonson as writer/artists, each of whom managed to combine comedy, adventure, and personal drama in a way that captured the spirit of Lee and Kirby.

Writer Tom DeFalco and artist Paul Ryan have inventively sought to bring the Fantastic Four into the nineties, setting the Torch at odds with the law, making Sue more aggressive (and giving her a far more revealing costume), scarring Ben's face, and even writing out Reed for a time. They also dispatched Reed and Sue's son Franklin into the future, from which he returned as Psi-Lord, an adolescent with mental powers and the founder of a spin-off team, the Fantastic Force. In the more violent Marvel Universe of Venom and Cable, the Fantastic Four reflects a more innocent, optimistic time. Indeed, the *Marvels* series presented the wedding of Reed and Sue as the shining culmination of a new Golden Age of heroes; after that the darkness began to spread.

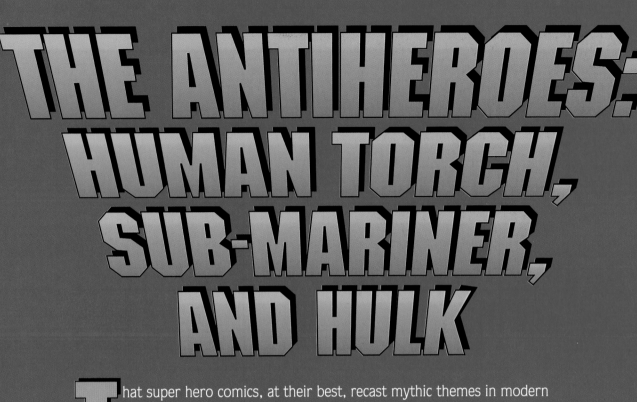

THE ANTIHEROES: HUMAN TORCH, SUB-MARINER, AND HULK

That super hero comics, at their best, recast mythic themes in modern pop cultural trappings is especially clear in the very beginnings of the genre. The great characters of the 1930s had a primal quality often lacked by those that followed. Superman (1938) was a sky god, who had descended to Earth from the heavens and traveled through them; the horned Batman (1939), associated with night, was like a demon turned to the service of good, emerging from the earth (the Batcave) to fight evil. Marvel's first super heroes, the Human Torch (fire) and the Sub-Mariner (water), debuting in *Marvel Comics* #1 in 1939, completed this set of the four alchemic elements and hence—as Marvel soon discovered—made natural adversaries.

The heroes of myth were not always devoted to the forces of good, and in this Marvel's first super heroes followed in their footsteps as well. Unlike the Human Torch and the Sub-Mariner, most comic super heroes were clearly both members and protectors of American society. Even Superman, who came from another planet, still looked human and had been thoroughly integrated into our society. The stars of *Marvel Comics* were not human beings at all, and not heroes as American readers of the 1930s might have understood the term, so much as antiheroes: powerful protagonists who lacked such conventionally heroic qualities as self-control, psychological balance, and adherence to a higher order of justice and morality. The creation of such characters was to become something of a Marvel specialty in the ensuing years, and none exemplified the type as much as the Incredible Hulk.

Marvel Comics #1 (1939) Script and art: Paul Gustavson

LIGHTING THE TORCH

Don't let his name fool you—the original Human Torch was an android, a robot in human form, the creature of Professor Phineas T. Horton. On first becoming conscious, the Torch found himself imprisoned in an airless transparent tube and placed on display at a press conference being held by his "father," who proceeded to demonstrate the sole "flaw" in his creation: when he introduced the smallest amount of oxygen into the tube, the android burst into flame and yet, strangely, was himself unharmed by it. Horrified, the press unanimously branded the android a menace; perhaps the real reason for their furor was an unwillingness to accept the idea of an artificially created form of life. Without a trace of empathy toward this newborn being, Horton buried the Torch alive, immuring the airtight tube in a pool of cement.

The tube proved not to be airtight after all. Eventually enough air seeped through so that the Torch could force his own "birth," exploding out from his cement tomb. At first he ran amok through New York City, unable to control his flame, spreading fire wherever he flew. Though apparently possessed with an adult consciousness and knowledge of English, the Torch did not understand his own abilities and initially acted like a terrified child, thinking he was burning himself alive. The populace was just as frightened as he was: this was no icon of physical perfection gliding serenely through the skies, but a glowing fireball in human form. No nor-

Invaders #18 (1977) Script: Roy Thomas /
Pencils: Frank Robbins / Inks: Frank Springer

PULP HERITAGE

Like other super heroes of the early Golden Age, the Human Torch quickly lost his dark edges once he became popular. The brutality of many of today's costumed vigilantes in comics, however, actually hearkens back to their forebears in the comics and pulp fiction of the 1930s. Take, for example, another character who debuted in *Marvel Comics #1*, the original Angel, a Batman-like crimefighter who in one scene coldly breaks a criminal's neck (top). During World War II Stan Lee and artist Jack Binder created the original Destroyer (seen above in a panel from the 1970s Marvel series *Invaders*, which revived many of the 1940s characters), a costumed resistance fighter who took no prisoners and hence anticipated such vigilantes willing to kill as today's Punisher.

TORCH ORIGIN

Marvel Comics #1 (1939) Art: Carl Burgos

Marvel Comics #1 (1939) Script and art: Carl Burgos

In the first *Human Torch* story Professor Phineas T. Horton presented his android creation to members of the press. Branding this new artificial man a monster, the press forced Horton to bury the Torch. But this elemental force could not be restrained for long: escaping, the confused and frightened Torch unintentionally wreaked havoc in New York City.

Marvel Comics #1 (1939) Script and art: Carl Burgos

Fantastic Four Special #4
(1966) Script: Stan Lee /
Pencils: Jack Kirby / Inks:
Joe Sinnott

Lee and Kirby revived the
original Human Torch as
the slave of the Fantastic
Four's longtime enemy the
Mad Thinker, a master of
computers and robotics,
paving the way for an
exciting battle between the
two Human Torches. When
the original Torch finally
sided with the FF, the
Thinker had him "killed."
But an android cannot
truly die, and John Byrne
finally resurrected him as a
sometime member of the
West Coast Avengers.

mal person could even touch the flaming Torch
without being killed.

As the Torch's origin story, by his true creator,
artist Carl Burgos, continued, the android fell into
the hands of criminals hoping to use him as a
weapon for arson. The Torch, however, swiftly mastered his powers and developed a moral sense,
turning against his captors. The true sign that he
had achieved psychological maturity came when he
rejected Horton, realizing that the professor too was
merely out to exploit him.

From then on the Torch became a crimefighter
and even joined the New York City police department in his secret identity of Jim Hammond. The
general public came to regard him wholeheartedly
as a hero, and, in the tradition of Golden Age super
heroes, the Torch even took on a kid sidekick, Toro,
a real human boy who turned out to possess powers identical to the Torch's.

Nevertheless, the Torch prefigured Marvel's later
android heroes, the Vision and Machine Man, and the
many contemporary mutant heroes whose noble spirits often contrast with their freakish appearances. No
matter how normal the Torch looked as Jim Hammond,
he was set apart from the rest of humanity by the
very circumstances of his "birth," and he looked far
from ordinarily human in his super hero persona.

TERROR FROM THE SEAS

Bill Everett's creation, Namor the Sub-Mariner, was
even more disturbing and charismatic than the

Human Torch. His story, like the Torch's, actually
began with his father, Captain Leonard McKenzie of
the *Oracle*, who set out on a scientific expedition to
explore the Antarctic in 1920. While using explosives to eliminate icebergs in his ship's path, he
unwittingly wreaked havoc on an undersea civilization, destroying its castles and killing scores of its
blue-skinned people. As originally depicted by
Everett, Emperor Thakorr and other males of this
undersea race looked something like fish that had
evolved into humanoid form, while the women
were far more human in appearance.

Thakorr dispatched his own daughter, Princess
Fen, to spy on the invaders: "Thou shalt find thy
way into the hands of these white monsters, there
to work your feminine wiles to our racial advantage." She was soon captured by the *Oracle*'s crew,
but the captain and the beautiful undersea princess
fell in love and were married on shipboard. A party
of warriors, seeking the lost princess, boarded the
ship and shot McKenzie. (He was believed to be
dead but turned up decades later, only to be killed
by one of Namor's foes.) The ship departed, slaughtering more of Fen's people.

Safely back in her kingdom, Fen gave birth to
Namor, the first hybrid child of the human and
water-breathing species, who was able to breathe in
air for more than a few hours, as well as water. As
he grew older Namor proved to have superhuman
strength, and he even developed tiny wings on his
heels that somehow enabled him to fly (as later
writers pointed out, Namor was not only a hybrid,

Marvel Comics #1 (1939) Script and art: Bill Everett

The scene in the first Sub-Mariner story that defined the character's purpose took place in the fishlike emperor's throne room (right). The young Namor brings the creatures he has impetuously killed to his grandfather, only to discover that they are surface human beings in diving suits. He is nevertheless congratulated by his mother, the beauteous Princess Fen, who is bent on war with the surface world. *Marvel Comics* #1 did not have splash pages (as the full-page, decorative panels in comic books are called), but it did include a smaller panel displaying the teenage Sub-Mariner in his youthful glory (above).

Marvel Comics #1 (1939) Script and art: Bill Everett

THE DESTRUCTION
OF NEW YORK CITY

In the first crossover between two Marvel series, the Human Torch tried to stop the Sub-Mariner's reign of terror in the June 1940 issue of *Marvel Mystery Comics* (as *Marvel Comics* had been renamed with its second issue). This battle of titans proved tremendously popular with the readers and built to a dramatic peak with the sixty-page story "The World Faces Destruction" in *Human Torch* #5 (Fall, 1941), a title clearly reflecting American jitters on the eve of entering World War II, that climaxes when the Sub-Mariner and his undersea army launch a tidal wave at Manhattan, "so terrific it slams down the world's most famous skyline as if it were built with cards, and then, its fury still unspent, spans the Hudson River and roars westward! Goodbye Broadway! So long, Times Square! Down goes the Empire State Building! Down goes the George Washington Bridge!" In the original story, the scripter then tells us that "the spirit of the populace stays up" since they had all taken refuge in watertight underground shelters (almost certainly inspired by the air-raid shelters of London: Hitler's bombing campaign had begun that September).

The first issue of Kurt Busiek and Alex Ross's 1994 series *Marvels* deconstructs this very same story by showing it from the point of view of the besieged New Yorkers. To them the Torch and Namor are not glamorous, heroic figures but dangerous freaks whose personal war imperil them all; the tidal wave that sweeps Manhattan is a vision of the end of the world brought on by mythic figures who incarnate the forces of nature.

Marvels #1 (1994) Script: Kurt Busiek / Art: Alex Ross

Marvels #1 (1994) Script: Kurt Busiek / Art: Alex Ross

but also Marvel's first mutant hero).

One day the impetuous, quick-tempered adolescent stumbled upon an underwater salvage expedition and, mistaking its members for robots on account of their diving suits, brutally slew them and wrecked their ship. On bringing the divers' bodies before the emperor, Namor discovered to his surprise that they were humans after all, but he evinced no guilt or regret for taking their lives. Fen sent him to "the land of the white people," and Namor soon launched a one-man campaign of terrorism, wreaking destruction throughout the surface world in story after story, with New York City as his particular target. The rising of the Sub-Mariner's "blue" people against "white" America to avenge an ancient crime of genocide was an easily grasped metaphor for struggles in the nation's past, and somehow Namor did not seem an outright villain; readers understood that he abided by his own moral code, according to which he was a lone avenger and defender of his people.

Like Tarzan, Namor was a "natural" man, a "noble savage," unconfined by the rules and customs of Western society. In the late 1930s, his near nudity would itself have been considered shocking. Although Everett was inspired by the legends of sunken Atlantis in creating his undersea kingdom, he did not name it Atlantis and he set it in the Antarctic rather than in Atlantis's traditional location far to the north; he called his water-breathing race "sub-mariners" (after Samuel Taylor Coleridge's poem *The Rime of the Ancient Mariner*) rather than

Atlanteans. Still, he clearly had classical motifs in mind: Namor's name is "Roman" spelled backward. His pointed ears and exaggerated, hornlike eyebrows and hairstyle gave him an impish appearance, reminiscent of Pan, the wild god who foments chaos in civilization. (Stan Lee was alert to these mythological associations, and when he and Jack Kirby revived Namor in the 1960s it was as a worshiper of Neptune, whose trumpet summons monsters.) Moreover, the Namor of the late 1930s was still a teenager. One would not have expected to find so powerful an expression of adolescent rage and violence in the popular culture of the 1930s; Namor's rampages make the gunfights of some of today's costumed vigilantes in comics seem puny and restrained.

Everett soon realized that his outlaw hero had a romantic appeal and had him encounter Betty Dean, a New York City policewoman (and a surprisingly liberated heroine for the comics of that time) who single-handedly tried to arrest the Sub-Mariner. Playing Beauty to his Beast, she provided a moderating influence on his behavior; needless to say, she also turned out to be his love interest.

Inevitably, as time went on, the Sub-Mariner's behavior mellowed, and he turned to battling "acceptable" menaces like the Nazis rather than society itself. Namor joined with his former enemy the Human Torch and, strange as it would have seemed years before, America's greatest champion, Captain America, in forming the All-Winners Squad in the late 1940s, and by the 1950s Namor was yet

Fantastic Four #4 (1962)
Script: Stan Lee / Pencils: Jack Kirby / Inks: Sol Brodsky

In a Bowery flophouse Johnny Storm shaves a homeless amnesiac only to reveal him as the long-missing Sub-Mariner. In a wink to the audience, Lee and Kirby had Johnny recognize him from having read an old comic book!

Fantastic Four #4 (1962)
Script: Stan Lee / Pencils: Jack Kirby / Inks: Sol Brodsky

Using the Horn of Proteus, Namor summoned a Brobdingnagian sea creature (somewhat resembling Monstro the whale in Disney's Pinocchio) that lay waste to Manhattan until the Thing killed the beast.

Fantastic Four #14 (1963)
Script: Stan Lee / Pencils: Jack Kirby / Inks: Dick Ayers

Lee and Kirby had no qualms about concocting imaginary machines or creatures to meet their stories' needs. Bottom, Namor sends Sue Storm telepathic messages via "the wondrous mento-fish," which is, to noboody's surprise, described as "the only fish of its type in the world."

another New York City–based crimefighter.

Tremendous as the enthusiasm for super heroes had been in America in the early 1940s, it had almost completely faded by the end of the decade, when the Human Torch, the Sub-Mariner, and Captain America ended their first run. They surfaced briefly in the comics of the 1950s, only to vanish again until the Marvel super hero revival in 1961. Of the three, the Torch has been seen the least since then, largely because Stan Lee and Jack Kirby created a new Human Torch—who really was human—to serve in the Fantastic Four, but the original Human Torch continues to appear occasionally in his Jim Hammond identity, trying, as he puts it, to learn what it really means to be human.

THE SUB-MARINER REBORN

Lee and Kirby brought back the Sub-Mariner in *Fantastic Four* #4, inventively revealing that he had spent the years since his last appearance as an amnesiac homeless man on the streets of Manhattan. Johnny Storm found and recognized Namor and immersed him in water, thereby restoring his memory and full powers. Namor soon learned that his people's undersea homeland had been devastated and deserted and he—wrongly, as it turned out—blamed the humans of the surface world. With this one stroke Lee and Kirby restored Namor to his most dramatic role: the avenger at odds with the human race. His innate nobility reemerged in the next issue, in which Namor sided

with his erstwhile enemies against the greater threat of Doctor Doom. Lee and Kirby also recognized the Sub-Mariner's romantic appeal, and the Fantastic Four's Sue Storm found herself torn between her attraction to her conventional boyfriend Reed Richards and Namor's more exciting, illicit appeal.

Once Lee gave the Sub-Mariner his own series in *Tales to Astonish*, written by himself and drawn by Gene Colan, Namor's role underwent a considerable

ATLANTIS

Namor found his lost people in the first *Fantastic Four Annual* and led them in a full-scale invasion of the major cities of the surface world, ultimately to be defeated by the stratagems of Reed Richards. Back in the 1940s, Everett had called Namor's people "sub-mariners," but they later acquired the name "Atlanteans." When Stan Lee and Jack Kirby revived the Sub-Mariner in the early 1960s, they had them return to the "original" site of Atlantis in the Atlantic Ocean and depicted both males and females as virtually identical to human beings save for their blue skin and pointed ears. (Still later, John Byrne hearkened back to Everett by giving the Atlanteans the black irises they had in their eyes in later Golden Age stories.)

Namor the Sub-Mariner #1 (1990)
Script and pencils: John Byrne /
Inks: Bob Wiacek

Fantastic Four Annual #1 (1963) Script: Stan Lee / Pencils: Jack Kirby / Inks: Dick Ayers

Sub-Mariner #8 (1968) Script: Roy Thomas /
Pencils: John Buscema / Inks: Dan Adkins

Perhaps the most affecting moment in Thomas's Serpent Crown epic was one of his characteristic nods to the comics of the past. Early on, the Sub-Mariner and Thing wreaked havoc in Manhattan fighting over possession of the deceased Destiny's helmet. A woman standing in the shadows caught the enraged Namor's attention and bid the startled Atlantean to leave the city peacefully. Not until the story's end was the city's savior revealed to be the now gray-haired Betty Dean, mourning the loss of her youth and her former Atlantean lover.

sea change, so to speak. Though he would still clash from time to time with the land-based super heroes, Namor was no longer primarily an outlaw. Rather, as prince of Atlantis he was now an authority figure, guarding his realm against threats from within and without.

Lee managed to get around Namor's new status as an establishment figure by concocting stories that removed Namor from his throne and dispatched him on solitary quests. Namor's *Tales to Astonish* series opened with a long, gripping story line about his quest to find the sacred trident of the Atlanteans' god Neptune in order to reclaim his throne from the usurping warlord Krang. In *Sub-Mariner*, writer Roy Thomas wove an even more intricate tale that spanned centuries and two oceans, ranging from modern-day America to a lost undersea civilization. He began by returning to the Sub-Mariner's origin story and adding to it: McKenzie, it turned out, had been accompanied on his expedition to Antarctica by a mentalist named Destiny who discovered there a strange helmet that greatly multiplied his psychic abilities. It was Destiny, who, decades later, was truly responsible for the destruction of the Atlanteans' Antarctic city, killing the Emperor Thakorr and (seemingly) Namor's mother, Fen (who, like McKenzie, turned up alive years later, only to die after her brief reunion with Namor), and who wiped out Namor's memory. Confronted by Namor and succumbing to his own growing madness, Destiny eventually hurled himself from the top of a Manhattan skyscraper. The helmet's story was not

yet finished. Brought to Atlantis by Namor, it proved to be the Serpent Crown, the source of the mystical might of Naga, the ruthless, serpent-faced tyrant of the undersea kingdom Lemuria and another antagonist for Namor.

Bill Everett returned to his creation as co-plotter and artist for several stories in the late 1960s and early 1970s. Everett's last major contribution to the Sub-Mariner mythos was the creation of Namorita, the spunky teenage daughter of Namor's late cousin, Namora, a character last seen in the *Sub-Mariner* stories of the 1950s. Everett died shortly thereafter, and the series was eventually canceled.

Sub-Mariner #11 (1969) Script: Roy Thomas / Pencils: Gene Colan / Inks: George Klein

Whereas Kirby's version of Namor was rather slender, as befits a swimmer, Gene Colan made the Sub-Mariner a more massively powerful, imposing figure, emphasizing his physical strength and regal presence. Stan Lee (and later Roy Thomas) complemented the new look by giving Namor a royal demeanor through his best pseudo-Shakespearean dialogue.

NAMORITA

Sub-Mariner #51 (1972) Plot and art: Bill Everett / Script: Mike Friedrich

Bill Everett's final major contribution to the Sub-Mariner mythos was his creation of Namorita, Namor's fiery-tempered but charmingly sweet young cousin (above left). After many years of absence from Marvel stories, Namorita proved popular when John Byrne brought her back in his *Namor* series, making her years older than the barely pubescent young girl Everett had depicted, but adhering closely to his characterization. With her youthful energy, high spirits, and familiarity with surface-world ways, Namorita provided an appealing contrast to the brooding, self-important Namor. Simultaneously, over in *The New Warriors*, writer Fabian Nicieza and artist Mark Bagley were casting Namorita as an aggressive, hard-bodied warrior (below). After Byrne's departure from *Namor*, *The New Warriors'* characterization won out. Namorita went through further mutation, emerging with Atlantean blue skin, webbed fingers, and a new name, Kymaera, that proclaimed she was no longer a diminutive version of her famous cousin (below left).

New Warriors #44 (1994)
Script: Fabian Nicieza /
Pencils: Darick Robertson /
Inks: Larry Mahlstedt

Namorita rises from the waves: *New Warriors #44 (1994)* Script: Fabian Nicieza / Pencils: Darick Robertson / Inks: Larry Mahlstedt

Fantastic Four #9 (1962)
Script: Stan Lee / Pencils:
Jack Kirby / Inks: Dick Ayers
If the Japanese can buy
American movie studios, so
can Atlanteans: Namor
turns up as producer of a
Fantastic Four movie, and
he's not dressed in his
usual swimwear.

When the Sub-Mariner finally won his own book again in 1990, it took a radically different form. In an early issue of *The Fantastic Four* Lee and Kirby had, in an unlikely turn of events, put Namor in a business suit and set him up as the head of his own Hollywood movie studio; the explanation was that Namor had access to undersea riches and had made considerable investments in the surface world. John Byrne, who disliked stories set in Atlantis, built upon this concept for his *Namor the Sub-Mariner*, in which the hero established himself as the head of a multinational corporation called Oracle. (Byrne explained away Namor's rampages against humanity in the stories of the early 1960s by revealing that he suffered from oxygen imbalances in his blood that sent him into manic-depressive fits.) Once again, Namor was not an outsider but an authority figure; in fact, he seemed rather stuffy and conservative compared to his teenage costar Namorita, whose lively characterization in the hands of Byrne made her increasingly popular among readers.

The money-mad 1980s were over when Byrne turned Namor into a billionaire businessman, and this new role for Namor did not last long after Byrne's departure from the series; subsequent writers returned the Sub-Mariner to Atlantis, but the book lost its direction and was again canceled in 1995. Perhaps Namor has had difficulty sustaining a solo series from the sixties onward because he was not meant to be the authority figure contemporary writers keep making him. The Sub-Mariner proved most successful in gripping the readers' imagination when he was an outlaw attacking conventional society. He is the true forebear of all of the rebels in today's comics. Cable, X-Force, and the Punisher—all of them follow the path that Namor the Sub-Mariner first blazed in 1939.

ENTER THE HULK

The Incredible Hulk, created by Stan Lee and Jack Kirby in 1962, succumbed to cancellation only six issues after he first appeared, but was soon revived and has been popular ever since. The Hulk provided a greater twist on the basic super hero concept than any other creation of Lee's. Whereas Superman pretended to be the meek Clark Kent, the Hulk's alter ego, Dr. Bruce Banner, was Clark Kent for real. Kent becoming Superman is the expression of a primal power fantasy: within the unassuming everyman is a being of godlike might and nobility that serves the good of all humanity. The Hulk series says something quite different: within everyman lies great, destructive power propelled by uncontrollable rage and egotism that can destroy civilization and overwhelm everything noble in the human spirit.

The Bruce Banner introduced in the opening pages of *Hulk* #1 is a darker variation on the familiar protagonist of Stan Lee's super hero stories: a Reed Richards with a dormant conscience. Thin, bespectacled, introverted, and narrowly focused on his work as a nuclear physicist, Banner had invented the "G-Bomb"—a "dirty" atomic weapon that

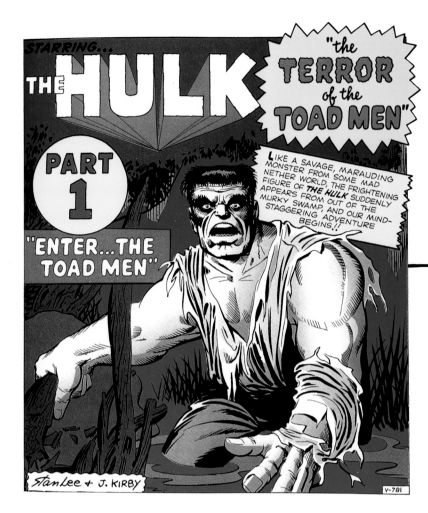

STARRING... THE **HULK**

"the **TERROR** of the **TOAD MEN**"

PART 1

"ENTER...THE TOAD MEN"

LIKE A SAVAGE, MARAUDING MONSTER FROM SOME MAD NETHER WORLD, THE FRIGHTENING FIGURE OF *THE HULK* SUDDENLY APPEARS FROM OUT OF THE MURKY SWAMP, AND OUR MIND-STAGGERING ADVENTURE BEGINS.!!

Stan Lee + J. Kirby

V-781

Incredible Hulk #2 (1962)
Script: Stan Lee / Pencils: Jack Kirby / Inks: Steve Ditko

Kirby's early Hulk bore a clear relationship to the various monsters that he had designed for Marvel stories just before the revival of super heroes in the 1960s.

released enormous amounts of lethal gamma radiation, much like the neutron bomb developed in real life years later—and is conducting a test at a base in a desert in the American Southwest. In charge of base security is Air Force General Thaddeus E. "Thunderbolt" Ross, whose name conjures up images of a wrathful Zeus. Ross, fiery-tempered and fiercely macho, loathes Banner: "The trouble with *you* is you're a *milksop!* You've got no *guts!*" What makes matters worse is that Ross's daughter Betty, another introvert, clearly under the thumb of her overprotective father, is attracted to the scientist.

Banner has a more serious enemy at the base, a Soviet spy named Igor employed as a member of Banner's scientific team. Much like Victor Von Doom when he first met Reed Richards, Banner is sufficiently arrogant to refuse to let Igor or anyone else double-check his formulas for the bomb: "I don't make mistakes, Igor!" Readers of early 1960s popular fiction presumably did not need to be told

(although later Marvel stories would point it out) that Banner's hubris has led him to create a monstrous weapon without considering the moral consequences. Banner suffers from a total lack of self-knowledge: when Igor, worried that something might go dangerously wrong with the test, seizes Banner by the lapels, Banner replies calmly, "You know how I detest men who think with their fists." And then he turns to watch the test that has been his dream, a display of the ultimate destructive power man can achieve. In fact Banner loves power, as long as he can project it into something distinct from himself, something that is the product of his intellect and not his emotions or his physical self—like the G-bomb.

The countdown to the bomb test is in progress when Banner learns that someone has driven onto the test site. Ordering Igor to halt the countdown, Banner goes himself to escort the intruder, a teenager named Rick Jones, to safety. It is a surprisingly selfless act for the seemingly amoral man we have been watching, but it comes too late. Seeing his opportunity to dispose of Banner, Igor allows the countdown to go forward. As soon as Banner has pushed Jones into a trench shielded against radiation, the bomb is detonated. Banner is irradiated with what should have been a lethal dose of gamma radiation but does not die. As in so many science-fiction movies of the 1950s and 1960s (and other

Incredible Hulk #1 (1962) Script: Stan Lee / Pencils: Jack Kirby / Inks: Paul Reinman

Mild-mannered physicist Bruce Banner quietly endured the continual blustering of General "Thunderbolt" Ross at the atomic testing site where they worked. Outwardly demure, Ross's daughter Betty even then had a will of her own and felt attracted to the inner strength of the man her father saw only as a weakling. In fact, Banner was developing the gamma bomb, a far-more-powerful weapon than even General Ross could imagine, and within Banner himself lay a rage that would make Ross's posturing look empty by comparison. Above, the bomb detonates and events quickly get out of hand.

Tales to Astonish #90 (1967) Script: Stan Lee / Art: Gil Kane

Like a very young child, the Hulk in Lee's *Tales to Astonish* would throw tantrums at the least provocation.

Marvel comics), nuclear radiation is depicted as a nightmare that can produce greater horrors than death.

Alone with the grateful Jones in the base hospital the following evening, Banner transforms into an enormous, superhumanly strong, grotesque monster; his personality changes too, becoming brutish, angry, contemptuous of anyone physically weaker than himself. Banner has become the very thing he denied in himself: the creature who thinks with his fists, what he later calls "that brutal, bestial mockery of a human . . . which despises reason and worships power!" The monster breaks free from the hospital and is sighted by a soldier, who commands, "Fan out men! We've got to find that—that Hulk!"

Stan Lee has recalled that the Hulk was inspired by the success of the Fantastic Four's own grotesque, bad-tempered monster, the Thing. But whereas the Thing was a part of the Fantastic Four's family unit and worked within society, the Hulk would be a true misfit, an outsider who could never belong. Boris Karloff's version of Frankenstein's monster was another inspiration: the Hulk was to be a misunderstood, persecuted monster for whom readers could feel sympathy. But while Banner, tormented by the curse of his nightly transformations into the beast, immediately became a sympathetic figure, the Hulk initially inspired only dread. The third inspiration for the Hulk, after all, was Dr. Jekyll and Mr. Hyde. Indeed, the Hulk was originally downright evil.

Thus was born a super hero whose underlying psychological makeup provided various writers and artists over the years with an amazingly rich ground

for developing nuances of characterization (extending even to an ongoing argument over whether the Hulk should be portrayed as green or gray). Originally, it was only Banner's personality that changed when he became the Hulk, not his intellect. Perhaps this reflects a difference in the attitudes of the Hulk's creators: the Hulk was smart in the stories drawn and co-plotted by Jack Kirby, both in his own series and in *The Avengers*. After the original *Hulk* comic was canceled, the Hulk was revived a year later as the second feature in *Tales to Astonish*, still scripted by Stan Lee but now drawn by Steve Ditko. In *Astonish* Lee showed Banner's intelligence sharply declining to the level of a small child's when he became the Hulk. In personality the Hulk was also like a child in a titanically powerful adult body.

If the Hulk continued to revel in his own power and easily flew into tantrums, he no longer plotted evil against the human race; now Lee's conception of him as a modern-day version of Karloff's monster

Incredible Hulk #147 (1972) Script: Roy Thomas / Pencils: Herb Trimpe / Inks: John Severin

In a remarkable 1970s tale reminiscent of a *Twilight Zone* story, the lonely Hulk once hallucinated an idyllic small town in which he found the thing he most desired: acceptance.

A HULK OF A DIFFERENT COLOR

Incredible Hulk #375 (1990) Script: Peter David / Pencils: Dale Keown / Inks: Bob McLeod

Incredible Hulk #377 (1991) Script: Peter David / Pencils: Dale Keown / Inks: Bob McLeod

Stan Lee gave the Hulk gray skin in his first issue, thinking the color would fit the somber mood of the series. Deciding it had been a mistake, Lee changed the Hulk's color to its more familiar, eerier green hue in issue #2; for the next two decades reprintings and retellings of the Hulk's origin made him green from the beginning. But in the 1980s *Hulk* writer/artist John Byrne established that the Hulk "really" was gray in his origin, and writer/artist Al Milgrom, who followed Byrne briefly on the series, shook up reader expectations by turning the Hulk back into his original gray-skinned form. Thus even a "mistake" in what becomes regarded as a classic story can take on the authority akin to holy writ for later generations of fans.

In these panels writer Peter David and artist Dale Keown symbolically portray the gray and green Hulks as separate personalities within Banner's fragmented psyche. Also pictured (right) is Banner's psychiatrist, Doc Samson, who himself was endowed with superhuman strength (and green hair) by gamma radiation.

fully took hold. Lonely and longing for friendship, especially from General Ross's daughter Betty, all he wanted from the larger world was to be left alone, free from persecution. The character's dual nature provided all the conflict the series needed: Banner was the overly civilized adult who had lost touch with his emotions; the Hulk was what would years later be termed the inner child that repeatedly burst forth with overwhelming strength. In the very first Hulk stories Banner turned into the monster at sunset, releasing conventional associations with night, sleep, and the unleashing of subconscious impulses in dreams. As the series went on, Banner's transformations instead came about whenever he became excited, as if this pathologically inhibited man could not indulge in the least emotion without risking loss of control.

The Hulk's adventures took over all of *Tales to Astonish*, which was renamed *The Incredible Hulk* in 1968. For years before this and for many more afterward, story lines revolved around the basic formula of a blending of Beauty and the Beast and a phantasmagorical Freudian triangle, with the stern, overbearing father, General Ross, dispatching entire armies to destroy the beast-man in love with his daughter.

As the Hulk evolved into a being who only became violent when provoked, his original Hyde-like capacity for evil was projected into two other green-skinned characters who became his foremost nemeses. The first was the Leader, who, in a clever twist on Banner's story, had been an ordinary blue-

Incredible Hulk #3 (1962)
Script: Stan Lee / Pencils: Jack Kirby / Inks: Dick Ayers

Incredible Hulk #153 (1972) Script: Gary Friedrich / Pencils: Dick Ayers and Herb Trimpe / Inks: John Severin

IMAGES OF REPRESSION

Repeatedly in the early Hulk stories we see the Hulk in a prison that Banner designed for his other self, pounding furiously on the walls, intent on bursting free (top); it is a striking visual icon for the repression of emotion.

The government forces seeking to capture and destroy the Hulk have often seemed worse menaces than he. Gary Friedrich's 1972 story "The World, My Jury!" makes explicit references to the infamous Chicago Seven trial in which radical defendant Bobby Seale was bound and gagged in the courtroom.

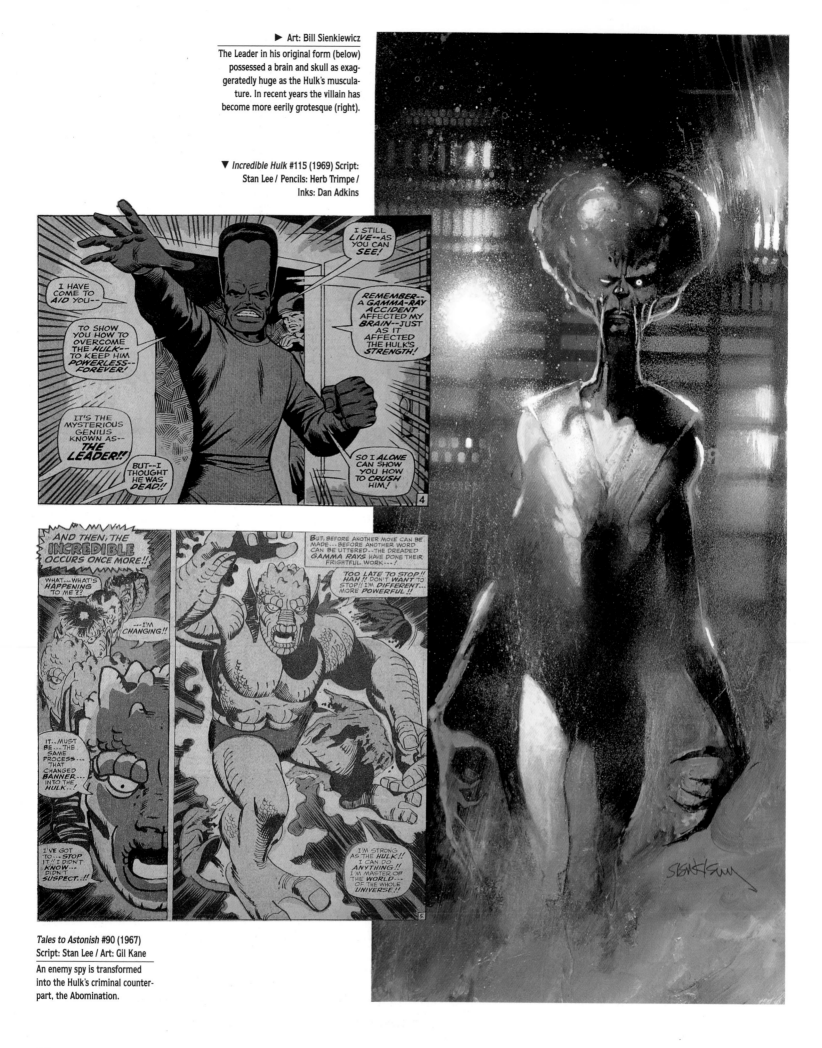

► Art: Bill Sienkiewicz

The Leader in his original form (below) possessed a brain and skull as exaggeratedly huge as the Hulk's musculature. In recent years the villain has become more eerily grotesque (right).

▼ *Incredible Hulk* #115 (1969) Script: Stan Lee / Pencils: Herb Trimpe / Inks: Dan Adkins

Tales to Astonish #90 (1967) Script: Stan Lee / Art: Gil Kane

An enemy spy is transformed into the Hulk's criminal counterpart, the Abomination.

collar worker transformed into a genius by gamma radiation (which had caused his brain and skull to expand dramatically) and who plotted world conquest. In the numerous struggles between the two enemies, the Hulk's low intellect, strong emotion, and physicality were pitted against the Leader's evil, cerebral cunning and disdain for those of lesser intelligence. However, it appeared that the Hulk's innocent stupidity and forceful emotion were more than a match for evil, cold intelligence.

The second was the Abomination, who had been yet another Communist spy before he became a grotesque gamma-irradiated monster. His strength rivaled the Hulk's, but he retained his normal intelligence and used his great powers for criminal ends. The Hulk, fitting the archetype of the primal "wild" man, had an innocence uncorrupted by civilization's temptations and inevitably won out over the Abomination as well.

A MIND DIVIDED AGAINST ITSELF

Clearly, Hulk was an example of multiple personality disorder, only in the Hulk's case the personalities also took disparate physical forms. Interestingly, just as Banner tried to deny and repress the passions and capacity for monstrousness within himself, so too, in his Hulk personality, he tried to repress Banner. For most of his history the Hulk thought of Banner as a different person, his greatest enemy whom he was never able to find or defeat.

Strangely, though, for more than twenty years

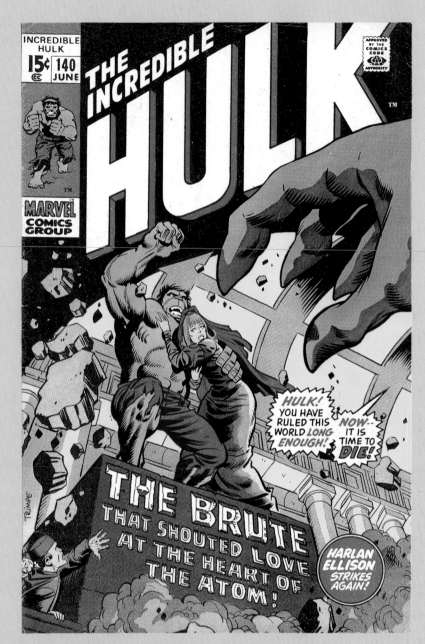

Incredible Hulk #140 (1971) Art: Herb Trimpe

PARADISE LOST

Author Harlan Ellison, a longtime comics enthusiast, created a paradise for the Hulk for a story turned into comics form by Roy Thomas and artist Herb Trimpe. Transported to a "subatomic universe," the Hulk found himself in the land of K'ai, where magic enabled him to retain Banner's intelligence and personality. There the Hulk became a heroic warrior and the true love of the beautiful Jarella, princess of a civilization of people whose skin was the same green hue as his own. Time and again over the years the Hulk returned to this paradise, only to be forcibly removed once more to the "real" world, where he fell victim to his own bestiality once again. Jarella was eventually killed off, but Ellison's vision of Banner and the Hulk integrated into one being anticipated later developments in the series.

Writer Bill Mantlo established
that Bruce Banner's father
went into uncontrollable rages,
physically lashing out at him
and calling him a "monster."
So it was that years later
Banner's psyche manifested his
fear and hatred in the form of
a literal monster, likewise
given to violent fury.

the Hulk's writers failed to demonstrate
any psychological connection between
Bruce Banner and the Hulk. Writer Doug
Moench, in the Hulk's short-lived magazine series, introduced a psychologist
who diagnosed Banner as a victim of
multiple personality disorder, but this
story was ignored. Roger Stern and
Peter Gillis even collaborated on a story
purporting to prove that the Hulk was an
entirely separate entity from Banner that
shared his physical form.

Finally, in 1985, in issue #312,
writer Bill Mantlo clearly defined the
psychological relationship between the
Hulk and Banner in a way that set the
stage for the most recent treatment of
the character. Mantlo devised a series of
flashbacks ranging throughout Banner's
life up to the events of *The Incredible
Hulk* #1, showing the child Bruce as the
victim of physical abuse by his hateful
father, who likewise tormented and finally killed
Bruce's mother. Bruce spent the rest of his childhood, his adolescence, and his early adulthood
internalizing and repressing his anger toward his
father and other oppressive figures, culminating
with General Ross himself. It was thus no accident
that when the gamma bomb explosion ripped open
the layers of repression in Banner's mind, it was the
monstrous fury of the Hulk that leapt to the surface.

Following Mantlo's stint John Byrne took over

Not such a nice guy after all: On taking over the
series, John Byrne immediately distanced his
Hulk from the sentimentalized, childlike Hulk of
the previous decade, who actually fondly referred
to deer as "Bambi."

Incredible Hulk #319 (1986) Script and art: John Byrne / Background inks: Keith Williams

Having temporarily divided the Hulk and Banner into separate beings, John Byrne brought Bruce and Betty together in a wedding ceremony that marvelously resolved many long-running story lines. As the Hulk rages outside, held off by Doc Samson, the uninvited "Thunderbolt" Ross (right), dressed in rags and tottering on the brink of his sanity, arrives to shatter the Freudian triangle once and for all by shooting Banner dead. Rick Jones intercepts the bullet, thus paying Bruce back for saving his life so long ago; luckily, he is only wounded (left). Meanwhile, Betty summons the courage to face down the father who had dominated her for so long. For a moment, Bruce and Betty put to rest the demons that had so long haunted them.

for an all-too-brief run as the Hulk's writer and artist. Byrne reemphasized the Jekyll-and-Hyde nature of the character by temporarily splitting the Hulk and Banner apart into separate physical beings. (This trick had first been performed by Roy Thomas in the 1970s, but merely for two issues.) Freed completely of the mild-mannered, intellectual Banner, the Hulk turned into the embodiment of irrational, unrestrained rage, now incapable of speech. As he did in *Namor*, Byrne revitalized the female lead in the series, in this case the spirited, independent Betty Ross, who became the series' true heroine over the subsequent decade.

Writer/artist Allen Milgrom, who followed Byrne, merged Banner with the Hulk once more. Peter David, who took over the series in 1987, has now authored well over a hundred issues, providing the most inventive stories for the character since his creation. Throughout his run David has repeatedly taken the principal characters in unexpected directions, exploring the essence of the series while shaking off the formulaic story lines of the past.

Following Mantlo's lead, David acknowledged Banner and the Hulk to be separate manifestations of his lead character's multiple personality disorder and was the first of the Hulk's writers to mount a sustained exploration of the psychological connections among the Hulk's personae. In fact, David even theorized that there were at least three separate personalities: Banner, inhibited but nonetheless capable of love and compassion; the gray, intelligent Hulk, who walled himself off from such emotions as

Incredible Hulk #349 (1988)
Script: Peter David / Pencils:
Jeff Purves / Inks: Terry Austin

One of the biggest shocks to the readership came when the Hulk seemingly perished in an atomic explosion, only to turn up as "Joe Fixit," a mob enforcer in Las Vegas, complete with a hat and suit. Nothing could seem more different from the uncivilized Hulk of the past, and yet this new role more clearly defined the Hyde-like aspect of the character: Peter David's Hulk was no longer a slow-witted innocent, but once again, the dark side of Bruce Banner's personality taken physical form, redeemed by his own code of loyalty to his associates.

a defense mechanism; and the traditional, green-skinned Hulk, an untamed child whose limitless rage manifested itself in nearly limitless destructive force. In one story line, psychiatrist Doc Samson succeeded in integrating these three personalities, producing what he believed to be a mentally and emotionally healthy and stable individual, who physically manifested himself as a handsome, green-skinned Hulk, with Banner's intelligence but without his repressions. Strangely enough, whereas in past decades the grotesquely overmuscled Hulk was meant to look like a monster, now his exaggerated physique matched that of various other 1990s heroes meant to embody a physical ideal. This new version of the Hulk briefly became the leader of the Pantheon, a band of superhumans who took it upon themselves to intervene in crisis situations on behalf of their view of justice, even if it meant opposing legal authorities.

In the story "Future Imperfect," by David and artist George Perez, the Hulk found himself transported to a possible postapocalyptic future, in which civilization had fallen, and the world was ruled by a bearded, green-skinned tyrant, the Maestro, who proved to be the Hulk's own future self. Unlike the Hulk of old, who threw fits of rage to demon-

Incredible Hulk #391 (1992)
Script: Peter David / Pencils: Dale
Keown / Inks: Mark Farmer

At times in the 1990s the Hulk surprisingly began to resemble Arnold Schwarzenegger's action-adventure movie characters.

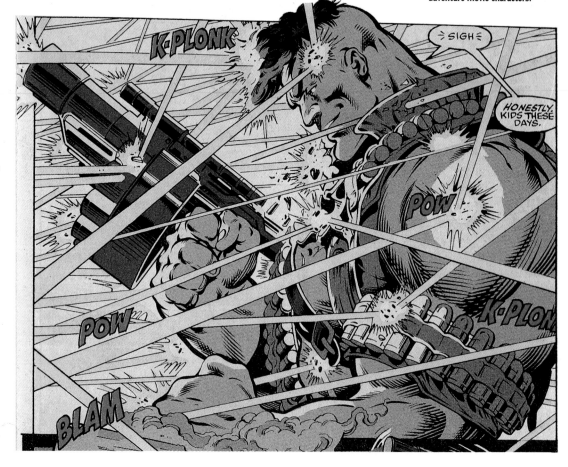

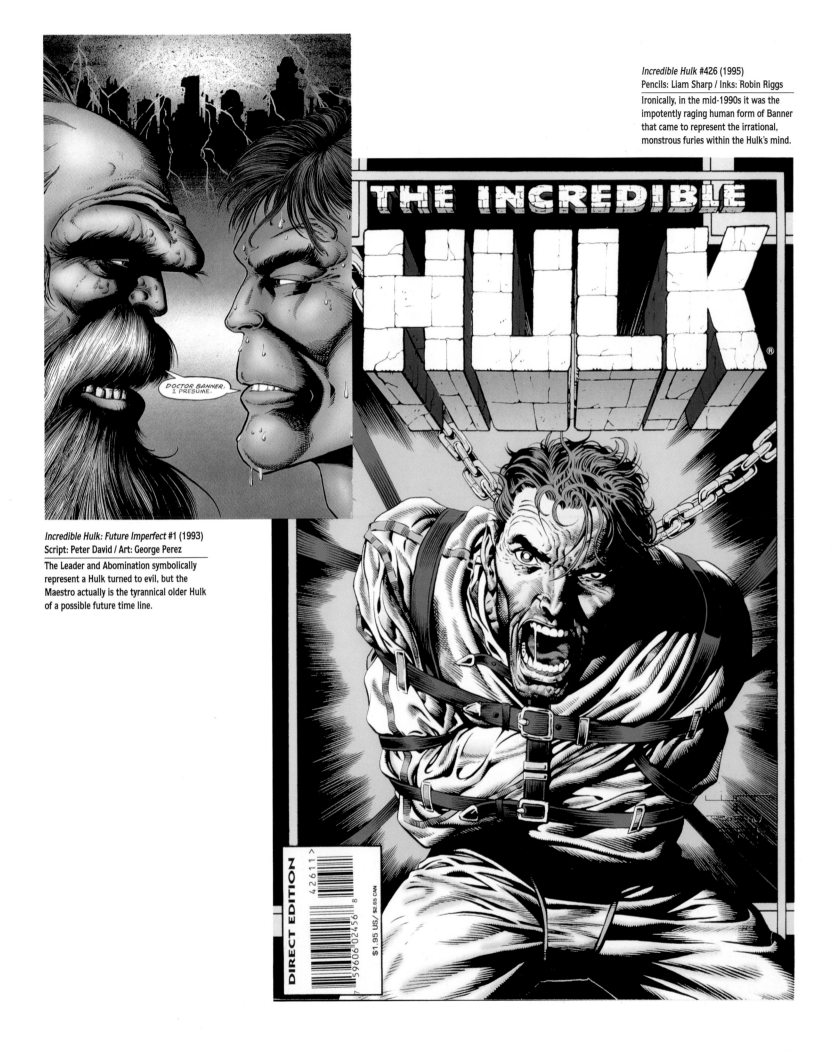

Incredible Hulk #426 (1995)
Pencils: Liam Sharp / Inks: Robin Riggs
Ironically, in the mid-1990s it was the impotently raging human form of Banner that came to represent the irrational, monstrous furies within the Hulk's mind.

DOCTOR BANNER, I PRESUME.

Incredible Hulk: Future Imperfect #1 (1993)
Script: Peter David / Art: George Perez
The Leader and Abomination symbolically represent a Hulk turned to evil, but the Maestro actually is the tyrannical older Hulk of a possible future time line.

THE INCREDIBLE HULK

DIRECT EDITION

$1.95 US/ $2.65 CAN

Sensational She-Hulk Vol. 2 #4 (1989) Script and art: John Byrne / Inks: Bob Wiacek

SHE-HULK

Originally, back in 1980, Stan Lee and John Buscema created the She-Hulk as a virtual female duplicate of the male Hulk. Bruce Banner's cousin, the similarly quiet and somewhat mousy Los Angeles lawyer Jennifer Walters, was shot by a gangster. Banner gave Walters a transfusion of his own blood to save her life, which caused her to transform into the green-skinned, superhumanly strong "Savage She-Hulk." But the original *Savage She-Hulk* series proved to be a dead end for the character.

Then writer Roger Stern came up with a new take on the She-Hulk when he had her join the Avengers. Whereas becoming the Hulk released the repressed Bruce Banner's rage, becoming the She-Hulk turned the inhibited workaholic Jennifer Walters into a glamorous, sexy, fun-loving extrovert. She projected a joy in being a super hero that was rare among Marvel's angst-ridden characters. Moreover, whereas Banner and the savage Hulk each represents a different side of a fragmented personality, the human Jennifer is merely a phase that the emotionally liberated She-Hulk has outgrown.

John Byrne made the She-Hulk a temporary member of the Fantastic Four while the Thing was on leave. Byrne then gave the character an added twist when he started writing and illustrating *The Sensational She-Hulk* in 1989: now Jennifer was very much aware that she was in a comic book, would talk directly to the readers, commenting on the story, and even have arguments with Byrne, who would sometimes argue back in the captions. She would step over panel borders to change scenes and at one point even tore holes in panels in order to get from one page to another. In fact what Byrne had created in *The Sensational She-Hulk* was Marvel's first truly postmodernist comic book series.

Byrne used *She-Hulk* to satirize the conventions of super hero stories affectionately, to parody then current trends in comics (as with his team of dead mutant heroes, the X-Humed), and to ridicule the macho excesses of contemporary comics from Jennifer's perspective. But no matter how absurd the situations in which she found herself, Jennifer remained a beacon of common sense and good humor, the emotionally balanced center of a mad world.

strate his power, this future Hulk had followed a path like the Leader's, joining brute force with cruel cunning to master mankind. Even more disturbingly, the Hulk later recalled, "When I looked into the Maestro's face, I saw my father's eyes looking back."

Since returning from this time, the Hulk has feared giving way to the Maestro's madness. Yet another of his multiple personalities has now emerged in physical form: when the Hulk becomes angry, he transforms into Bruce Banner. Banner's scrawny form, animated by colossal rage, thus becomes a startling new symbol for repression: the hungry demon of the unconscious longing to devour the larger portion of the whole mind and body.

Removed from the Pantheon, the Hulk returned to his former role as wanderer, trying to escape persecution by the greater community. But whereas once he was an animalistic creature clad in rags leading a solitary existence in the desert, now he was trying to lead a life of obscurity in the small towns of America, disguised as a normal (if hugely built) man and accompanied by his wife Betty. Eventually, the military found him and he became a fugitive once again.

Betty Ross Banner has completed her transformation from the stereotypical devoted but frightened ingenue she was in the 1960s to a woman with the strength of will to stand up to the dangers of the Hulk's world and compassion and understanding sufficient even for dealing with her unpredictable mate. David's treatment of the relationship between the Hulk and Betty has become a strangely moving

Incredible Hulk #344 (1988) Script: Peter David /
Pencils: Todd McFarlane / Inks: Bob Wiacek
Betty succeeds in reaching out to her husband within
the gray Hulk's psyche. (Her pregnancy, which she
mentions, never came to term.) Todd McFarlane first
made a strong impact as an artist through his work on
the *Incredible Hulk* in the 1980s.

metaphor for any man and woman reaching out to
each other despite their own emotional scars. It is
portrayed in the sharp, witty dialogue that utilizes
irony to undercut the pretentiousness of the conventions of super hero action stories and is David's hall-
mark as a writer; his characters acknowledge the
absurdities of their world and seem more realistic for it.

None of the classic Marvel heroes have changed
more since the 1960s than the Hulk, and yet in
their essential qualities both the character and the
series have been consistent. Indeed, the Hulk stands
as a model of the conceptual strength of the greatest Marvel characters. Without having their pasts
discarded from the official canon of continuity,
these characters continue to evolve and to adapt to
the tastes of succeeding generations of writers,
artists, and readers, and, in the Hulk's case, remain
on the cutting edge of the super hero genre.

Hulk 2099 #3 (1995) Script: Gerard Jones / Pencils: Malcolm Davis / Inks: Chris Ivy

HULK 2099

The *Hulk 2099* series, scripted by screenwriter Gerard Jones, was set in a
nightmarish Hollywood of the future, where the wars between studios
can involve literal bloodshed. In transforming into the Hulk of 2099, a
particularly grotesque green-skinned monster (who has clearly inherited
the serpentine tongue of Spider-Man's adversary Venom), John Eisenhart
saw himself as becoming an archetypal noble savage, a wild man glory-
ing in his physicality and freedom of action, unconfined by the rules and
repressions of a society he believes has forfeited his allegiance.

THREE

YOUR FRIENDLY NEIGHBORHOOD SPIDER-MAN

Although by 1962 Stan Lee and Jack Kirby had thoroughly shaken up the super hero genre with their creations of the Fantastic Four and the Hulk, Marvel had yet to create a new super hero in the traditional sense of the term. Beginning in the late 1930s the super hero had been a person who adopted a new, costumed identity to wage war against evil while concealing his everyday identity and passing as an "ordinary" member of society. The Fantastic Four did not have secret identities, and they were more like adventurers and explorers than crimefighters; the Hulk was not a crusader for justice but a rampaging monster.

Lee and his collaborators had yet to show what they could do with the masked-crimefighter concept. They did it in *Amazing Fantasy* #15, cover-dated August 1962, less than a year after the Fantastic Four's first appearance, when the Amazing Spider-Man, created by Stan Lee and Steve Ditko, made his debut. The story goes that since *Amazing Fantasy* was being canceled with issue #15, Stan Lee decided to ignore commercial considerations and do the kind of super hero tale he really wanted to see done. Spider-Man was an experiment that was so different from the other adventure series of 1962 that not even its creators truly believed it would succeed.

But succeed and prosper it did: today Spider-Man is the symbol of the entire Marvel Entertainment Group, its most recognized character in the world at large. He stars in four monthly series—*Amazing Spider-Man*, *Spectacular Spider-Man*, *Web of Spider-Man*, and just plain *Spider-Man*—his own animated television

Untold Tales of Spider-Man #1 (1995) Script: Kurt Busiek / Pencils: Pat Olliffe / Inks: Al Vey

Only a high-school sophomore when he first became Spider-Man, the teenage Peter Parker already was weighed down by troubles on all sides. Kurt Busiek's *Untold Tales of Spider-Man* and *Amazing Fantasy* limited series tell new stories set in between *The Amazing Spider-Man* issues of the 1960s and superbly recapture the spirit of those classics.

series, and on dozens of different products. It is only right that Spider-Man has achieved such fame and popularity, for it was Spider-Man, the throwaway experiment, who revolutionized the entire super hero genre.

THE SPIDER AND HIS ENEMIES

In the early 1960s, when even the grim Batman of the thirties had given way to the grinning, square-jawed Caped Crusader familiar to all, Spider-Man's name alone signaled that something would be different about him. Indeed, Stan Lee was discouraged by his publisher, Martin Goodman, from naming his new character after as repellent a creature as a spider. Lee has stated that he was inspired by the Spider, a masked vigilante of 1930s pulp fiction. In evoking pulp thrillers Lee was clearly interested in setting his new hero in a more shadowy, dangerous milieu than the super hero genre of the 1960s was depicting.

Furthermore, Spider-Man concealed his features entirely. Even Batman's mask allowed onlookers to read his facial expression; Spider-Man's made his face a blank, save for the web pattern and huge, always staring blank eyes, somewhat reminiscent of a fly's. Artist Steve Ditko made him move like an insect, too, as he crawled up walls like a spider ascending its web or hung from the ceiling. From his wrists he shot sticky "webbing," which he employed to swing from one rooftop to another. (The webbing did not come from his body, as a real spider's did, but had been concocted by Spider-Man in a laboratory.)

Spider-Man's "web-swinging" is such a familiar sight to comics aficionados now that they have surely forgotten just how weird it looks at first glance.

Like the vigilantes of the pulps, notably the Shadow, Spider-Man often operated at night. Stalking criminals, terrifying them by catching them within the spotlight cast from his belt (complete with its own web pattern), Spider-Man would wrap them in his webbing, as a real spider does with its prey, and leave them for the police.

Instead of living in the all-too-familiar fictional metropolis (with either a lower- or uppercase "M") of super hero comics, Spider-Man made his home in a real-life New York City. Steve Ditko's New York ran the gamut from the aging, mundane *Daily Bugle* building to ruined warehouses, from dilapidated water towers atop roofs to the dingy sewers beneath the streets. Crime was not an unusual intrusion into the community's stability; it was everywhere. In the very first story it claimed the life of Spider-Man's uncle. Organized crime and gang wars have been a continual presence in Spider-Man's stories over the decades, beginning with Lee and Ditko's Big Man and Crime-Master. One of Spider-Man's greatest enemies has been Wilson Fisk, the Kingpin of Crime, a villain built like a sumo wrestler, powerful enough to hold his own in combat even with Spider-Man, who rules East Coast organized crime from behind his respectable businessman's facade.

If Spider-Man was eerie, then his principal costumed opponents were downright grotesque, as Lee and Ditko succeeded in fusing the costumed super-

Amazing Spider-Man #33 (1966) Plot and art: Steve Ditko / Script: Stan Lee

STEVE DITKO

Like Jack Kirby on *FF*, artist Steve Ditko seemed to expand his powers as *The Amazing Spider-Man* series took off. Perhaps the most memorable sequence in the character's history is this one, in which Spider-Man found himself pinned under tons of fallen machinery. As his struggle to free himself grew more intense, the panels grew larger, until finally Spider-Man heroically threw off the immense weight in a single full-page shot.

villain with memorably repellent villains in the Dick Tracy mold. The very first issue of *Amazing Spider-Man* introduced the Chameleon, the disguise master whose true visage still remains a mystery. The next issue brought with it the Vulture, whose winged costume enabled him to fly and whose bald head and beaklike nose made him the image of his avian namesake. The Vulture was an elderly man, empowered by his costume, a symbol of greedy Age and Death pitted against Youth and Life in the person of Spider-Man.

Then there was the Lizard, yet another of Lee's variations upon the Jekyll-Hyde theme. In a sense he was Spider-Man gone wrong: whereas Spider-Man was a science student who held onto his humanity in taking on his arachnid guise, Dr. Curt Connors was a scientist who lost his through his Faustian bargain with science. Normally he was a humane man with a loving wife and son, but in seeking to do away with his one "flaw," his missing arm, he unleashed the darker side of his personality. Drinking a serum that would give him a lizard's ability to regenerate severed limbs, he grew back his own arm but at the price of turning into the Lizard, a savage reptile with human intelligence. Time and again over the years Spider-Man has stopped this cold-blooded creature from slaying his own family. Rounding out the animal theme were Kraven the Hunter (who clothed himself in animal skins and sought to hunt down and kill Spider-Man as if he

SPIDER-MAN VILLAINS

THE KINGPIN, *Amazing Spider-Man #69* (1969) Art: John Romita, Sr.

Spider-Man may have the most colorful and memorable rogues' gallery in the whole Marvel canon. Like himself, many of his adversaries are human beings who have taken on the aspect and abilities of animals, including Doctor Octopus, the Vulture, and the Lizard, who is a human who actually transforms into a reptile. Even Kraven the Hunter wears a costume with animal motifs and hunts Spider-Man as if the hero were a beast, not a man. Another kind of shapeshifter is the Sandman, who can turn his body into rivers of flowing sand. The eerie costume of Mysterio, who employs his mastery of movie special effects for crime, would make him look at home in the surreal settings of Steve Ditko's *Doctor Strange*. With the Green Goblin, Lee and Ditko transplant a motif from medieval fantasy into a contemporary urban world.

THE VULTURE, *Amazing Spider-Man #2* (1963) Script: Stan Lee / Art: Steve Ditko

KRAVEN THE HUNTER, *Amazing Spider-Man #15 (1964)* Art: Steve Ditko

THE LIZARD, *Amazing Spider-Man #6 (1963) Script: Stan Lee / Art: Steve Ditko*

72

DOCTOR OCTOPUS, *Amazing Spider-Man* #11 (1964) Art: Steve Ditko

THE SANDMAN: *Amazing Spider-Man* #4 (1963)
Script: Stan Lee / Art: Steve Ditko

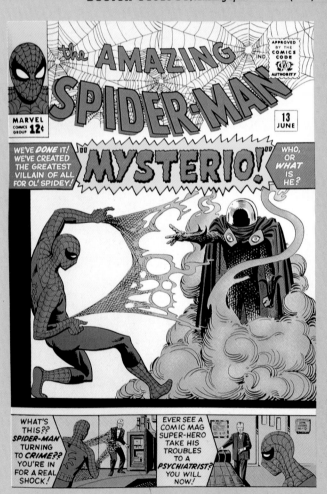

MYSTERIO, *Amazing Spider-Man* #13 (1964) Art: Steve Ditko

THE GREEN GOBLIN, *Amazing Spider-Man* #39 (1966) Art: John Romita, Sr.

Amazing Spider-Man #12
(1964) Script: Stan Lee /
Art: Steve Ditko

Of all super heroes, only Spider-Man could be humiliated this way. Fighting Doctor Octopus while ill, Spider-Man is easily defeated and unmasked. Yet neither Octopus nor Peter's boss J. Jonah Jameson nor even his girlfriend Betty Brant will believe that the studious milquetoast Parker could possibly be the heroic Spider-Man.

were not a human being at all), the Beetle, the Jackal, the Rhino, and the Scorpion.

Perhaps Spider-Man's greatest enemy of all has been Doctor Octopus, a pudgy, bespectacled nuclear scientist, originally withdrawn and gentle, who invented a harness with four robotic arms for handling radioactive materials. The radiation unleashed by an accident fused the harness to his body and distorted his mind, releasing a second personality, cunning and ruthless; worse, it created a psychic bond between the doctor and the robotic arms, enabling him to control them mentally. Over the years Octopus became Spider-Man's most persistent adversary, a grotesque figure whose serpentine, grasping metallic arms make him seem as much machine as man.

These men literally or figuratively turned into beasts defied the order of nature, but other villains in Spider-Man's rogues' gallery went even further. The Sandman was on the surface an ordinary thug, but yet another nuclear accident enabled him to

turn into sand, allowing him to shape-shift into flowing rivers of grains or fuse at will into a rock-hard living statue. This theme went further with the Green Goblin and his successor, the Hobgoblin, who seemed like medieval gargoyles come to life. Defying the laws of reality was the stock-in-trade of Mysterio, whose humanity was masked by his bubblelike helmet. He employed his genius at devising special effects to accomplish seemingly impossible feats, often with the goal of driving Spider-Man mad.

THE MAN BEHIND THE MASK

These villains bring out the "Spider" side of the character of Spider-Man: the avenger in a threatening, unstable, contemporary urban world. But it is the other half, the "Man," the very real humanity of the character, that has won him his vast, long lasting popularity. For one thing, when his series began, he was not a hardened crimefighter but Peter Parker, a fifteen-year-old high-school student sheltered from the potential cruelty of the world, a virtual innocent.

For decades there had been teenage costumed heroes in the comics, but they were almost always presented as junior partners to adult heroes, miniature versions of their elders, often given names (like "Robin the Boy Wonder") and costumes (like Robin's short pants) that emphasized their status as juveniles. The rationale was that the "kid sidekicks" were there as identification figures, but surely the young readers would prefer to be like the stars of the show, the adult heroes, who did not have to take orders. Marvel had created kid sidekicks in the Golden Age—Captain America's Bucky and the Human Torch's Toro—but Stan Lee disliked the whole kid sidekick gimmick and was determined to do something else.

In creating Spider-Man, Lee and Ditko placed a teenager at center stage as a lead character who had to act and make decisions on his own. In naming him "Spider-Man" and not "Spider-Boy" they signaled that they would give this character the same respect they would to an adult character. (The following year Lee and Kirby similarly dubbed their team of teenage mutants the X-Men, whereas National would end up calling its team of young heroes the Teen Titans.) Indeed, their greatest triumph in *Spider-Man* was their ability to present and examine the emotions of a maturing adolescent from the critical vantage point of older men, but without ever condescending to him or to his problems. By taking Peter Parker seriously, Lee and Ditko demonstrated that they were taking their readers seriously as well.

People often say glibly that Marvel succeeded by blending super hero adventure stories with soap opera. What Lee and Ditko actually did in *Amazing Spider-Man* was to make the series an ongoing novelistic chronicle of the lead character's life. Most super heroes had problems no more complex or relevant to their readers' lives than thwarting this month's bad guys or keeping their girlfriends from learning their secret identities. Peter Parker had far

more serious concerns in his life: coming to terms with the death of a loved one, falling in love for the first time, struggling to make a living, and undergoing crises of conscience.

ORIGIN IN REMORSE

The fact that Peter Parker was a super hero, operating in a world where making the wrong decision could be a matter of life or death, elevated his personal problems to an operatic scale. Spider-Man's origin is a fine example: it serves as a parable of how a youth crossing the borderline of maturity discovers his own potential to affect the world around him for good or ill.

The Peter Parker we first meet is desperately lonely and unhappy, shunned and held in contempt by his classmates, both male and female. Besides his love of science, the saving grace in his existence is his idyllic life with his surrogate parents, his Aunt May and Uncle Ben (Peter's parents died when he was a child). They provide him with a refuge in which he can be happy and which he naively assumes is secure from harm.

The change in Peter's life occurs through sheer accident. The death-and-rebirth motif in so many super hero origins is reduced here to something as mundane as Peter being bitten by a spider while visiting an atomic research lab on a school trip. The spider just happened to have been exposed to radiation; somehow the spider's bite mutates Peter, bestowing the powers of a spider on him. Just as

SPIDER-MAN ORIGIN

Amazing Fantasy #15 (1962) Script: Stan Lee / Art: Steve Ditko

The classic pattern of the hero's journey often begins with the protagonist in a lowly state, his virtues unrecognized by those around him. The early Peter Parker was shy, unathletic, and nerdish—an introverted bookworm. Above, he finds himself cast out by the rest of his high-school community, notably by big man on campus Flash Thompson, who later became Spider-Man's biggest fan. Beneath the pseudoscientific talk about radiation lies a mythic motif: Peter Parker undergoes a symbolic death and rebirth as he is bitten by a spider (top right), whose powers then enter into his body. Gaining his new powers releases a buried creativity in Peter, as he quickly devises weapons, a costume, and a new identity for himself (above right). The mythic hero errs when he refuses the call to adventure, which, in Spider-Man's case, was the opportunity to begin his crimefighting career by stopping an anonymous thief (right). Spider-Man realizes in horror the consequences of refusing the call when he recognizes his uncle's killer (below right). The justly famous final panel of the story states his credo (below left).

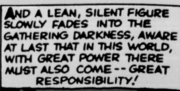

76

certain insects can lift many times their own weight, now Peter has superhuman strength. He can climb up walls like a spider by somehow sticking to them. He soon concocts the substance he will use as webbing. Strangely, he also discovers he has a form of extrasensory perception that alerts him to danger, which he dubs his "spider-sense," although what this has to do with spiders remains a mystery.

Peter's abrupt acquisition of super powers can be seen as his discovery of the inner strength and creative potential within himself. Now he realized he need not hide in the security of his childhood home from the cruel outside world, but could make his mark there and even dominate it. The spider's bite released a more aggressive, outgoing side of Peter's personality. Peter masked himself when using his powers in public, afraid at first of embarrassment if something went wrong. Soon the lonely introvert turned into an exhibitionist, seeking, even demanding acclaim for his superior abilities. Here is another of Lee and Ditko's innovations. What would someone in real life do who suddenly received super powers? On gaining their powers the Fantastic Four reverted to the convention of comics origin stories and immediately decided to become champions of humanity. Spider-Man took a far more realistic attitude, one particularly appropriate for the 1960s: he made himself into a media superstar, especially in the fast-growing, still relatively new medium of television. Peter actually created the costumed identity of Spider-Man as a stage persona, as a professional wrestler might. Hiding behind the mask of Spider-

Man ironically enabled the timid Parker to act entirely uninhibited before television audiences. Acclaimed by the public, Spider-Man's ego swelled.

Spider-Man carelessly stepped over a moral line one day when leaving a television studio after a performance. A guard was chasing a nondescript thief down a corridor, and Spider-Man just passively watched. The thief got away, and the guard demanded to know why Spider-Man did not help; the masked performer replied that it was none of his business.

Soon afterward, on his way home in his everyday identity, Peter learned that his uncle had been shot dead by a burglar. Anguished and infuriated, Parker changed into his Spider-Man costume and cornered and captured the terrified felon in his lair. To date Marvel has never given this burglar a name: his anonymity makes him representative of all criminals, even of the dark and destructive side of fate that cuts innocent people down. Already readers of the origin story must have been shocked, since it was conventional for any terrible fate that befell a member of a series' supporting cast to be reversed by the story's end. But Ben was dead, and there was another shock as well: his killer was the same thief whom Spider-Man had failed to stop at the television studio.

This double blow destroyed all of Peter Parker's illusions about himself and his world. A standard theme in super hero origins has the hero thrust into the role of crimefighter through the violent destruction of his sheltering family. The hero then seeks vengeance, but in the world of ongoing comics series, he can never get enough of it: the wounded hero focuses his fury on all evildoers in the world. When Spider-Man recognizes the burglar he had earlier failed to stop, however, he comes face to face with his own culpability. This is another reason that it seems right that the burglar is nameless: his function in Spider-Man's origin myth is that of a mirror. In fighting crime Spider-Man is not motivated by immature, inextinguishable rage against a world that has deprived him of emotional security. Nor is he motivated solely by altruism: Spider-Man is driven by his own guilt—he cannot allow anyone to come to harm if he can help it, and, thanks to his superhuman powers, he can. As Stan Lee intones at the end of the origin story, Peter had learned at a great price that "With great power must come great responsibility." This has remained the theme of the *Spider-Man* series for over thirty years. Over that time Spider-Man has repeatedly questioned his own mission, but he always comes back to the question of his own responsibility to the world. To allow wrongdoing to occur is, in Spider-Man's mind, to be part of it.

EVERYMAN IN TIGHTS

Until the coming of Spider-Man, super hero stories were almost always purely exercises in escapism. The reader could briefly forget his or her own troubles by vicariously participating in the hero's victory over his adversaries. Startlingly, Lee and Ditko placed Spider-Man in the same position as the

reader. Spider-Man inevitably triumphed in battle over his opponents, but he always returned to a troubled everyday life not very different than the reader's own. His superhuman powers could not solve these problems; in fact, his double life made them more complicated.

Despite his miraculous powers and glamorous new identity, Peter Parker also had to struggle to make a living; he too had to contend with strains and setbacks in his love life. The fact that he secretly led a double life as a super hero only magnified these everyday problems: a normal person might have to go to his office job while suffering from a cold; when Spider-Man got a cold in the middle of winter, he still felt obligated to go out to stop the Vulture. Mundane annoyances thus became entangled with matters of life and death.

Just as he had been shunned in high school, so too, as Spider-Man, he found himself isolated in the world at large, held in suspicion, and even feared, an icon representing the sense of alienation experienced by adolescents, a theme of growing importance in the culture of the 1960s. From the first issue of his own series Spider-Man has been pilloried in the pages of the New York City tabloid newspaper, *The Daily Bugle*, by its publisher, J. Jonah Jameson, as an irresponsible, glory-seeking vigilante and likely criminal. Jameson has turned much of the public against Spider-Man, and for years he was mistakenly wanted by the police for various crimes.

Furthermore, whereas most super hero adventures were fantasies about omnipotence (who was

Amazing Spider-Man #18 (1964)
Script: Stan Lee / Art: Steve Ditko

Unlike previous super heroes, Spider-Man was continually in need of money. Here, he is turned down by a trading-card publisher—an ironic turn of events indeed, given that comics trading cards eventually became enormously popular. In dealing with Spider-Man's real-life dilemmas, Lee and Ditko often found opportunities for comedy.

stronger than Superman or faster than the Flash?), Spider-Man, like many of the new Marvel heroes of the 1960s, was far from invincible. He might be stronger than an ordinary human criminal, but he was outmatched by opponents like Doctor Octopus and the Sandman. In fact, Jack Kirby was the original artist assigned to draw Spider-Man, but Stan Lee decided that Kirby made Spider-Man look too muscular, too conventionally heroic in build. So Lee assigned Ditko instead, who gave Spider-Man an ordinary, thin, even scrawny build that contrasted with his great strength.

Hence, the smaller Spider-Man really had to struggle to have any hope of overcoming his mightier foes. This disparity in power, of course, made the stories more dramatic, but also demonstrated that both in his "ordinary" life and in his super hero career, Peter Parker was continually contending against the limits of his capacity to affect his world and his life.

In short, Spider-Man is Everyman as super hero, beset as much by the banal miseries of everyday existence as by the grotesque super-villains who embody the forces arrayed against him. By establishing this dramatic contrast between the lead character's triumphs in battle and his sufferings in his inner life, *Spider-Man* set the pattern not only for the many Marvel series that followed, but also for

Amazing Spider-Man #50 (1967) Script: Stan Lee / Pencils: John Romita, Sr. / Inks: Mike Esposito (as "Mickey Demeo")

Over the years Peter Parker has repeatedly questioned the wisdom of continuing as Spider-Man, considering the havoc his dual role has wreaked on his personal life. At times he has even wondered if J. Jonah Jameson was right, that he is a potential menace. Invariably, these personal crises end when Peter reaffirms his dedication to Spider-Man's mission. Though being Spider-Man may seem an adolescent fantasy, the altruistic idealism that drives him has real moral value.

the entire super hero adventure genre as it evolved at many publishing companies over the next thirty years right through the present day.

Yet despite the grimness of Spider-Man's world and the tragedies in his life, the *Spider-Man* series is ultimately optimistic, even bright and cheerful in spirit. Although Spider-Man repeatedly sinks into melancholy, he is nevertheless a font of optimism and joy. If Spider-Man's wide-eyed, expressionless mask and animalistic poses make him look eerily inhuman, they also give him an appealingly cartoon-ish look. Other super heroes made jokes in the course of their adventures, but they were never par-ticularly clever: their banter seemed merely filler for word balloons during fight scenes. Spider-Man, however, had his (or, actually, Stan Lee's) own unique style of humor that became his trademark. His enemies invariably had grandiose egos and were forever pompously boasting of their own greatness in between threats to kill their comparatively insignificant foe. Spider-Man did not just win his fights with them physically; in the process he also cut their egos down to size with a constant barrage of wisecracks. They saw themselves as grand fig-ures of evil; he made them look ridiculous. He is the traditional little guy of comedy who outmaneuvers his enemies by outsmarting them.

THE SUPPORTING CAST

From the beginning the *Spider-Man* series has bal-anced its depictions of evil and suffering with cele-brations of those qualities that make the hardships of life easier to endure: love, friendship, and familial devotion. In no other Marvel series has there been so many noncostumed supporting cast members who have played such important and memorable roles in the hero's life.

One of the most colorful and prominent mem-bers of the supporting cast is J. Jonah Jameson. Jameson is Spider-Man's most implacable adversary, forever railing against him in print; Jameson has even recruited menaces like the Scorpion and Professor Spencer Smythe's Spider-Slayer robots to capture and unmask him. In one memorable sequence early in the series, Jameson confessed to himself that he tried so hard to tear down Spider-Man's reputation because he envied him, but he has never allowed himself so clear a look into his own motivations since then.

Aside from his irrational obsession with Spider-Man, Jameson is a sincere, committed idealist, campaigning for civil rights and against crime and showing genuine courage in standing up to pressure from criminals like the Kingpin. He is entirely dedi-cated to the *Bugle*'s staff, his second family, which, ironically, includes Peter Parker, who has derived his income all these years as a freelance news pho-tographer for the *Bugle*, using an automatic camera to take pictures of himself in action as Spider-Man. Somehow, all of Jameson's many sides fit together; the man who seems at first glance to be a mere car-icature proves to be surprisingly multidimensional.

If Uncle Ben was Peter Parker's "good" surro-

Amazing Spider-Man Annual #1 (1964)
Script: Stan Lee / Art: Steve Ditko

As early as 1964 Marvel ran feature pages on the workings of Spider-Man's powers and gadgets. Nearly twenty years later these features evolved into an encyclopedia of Marvel's characters, *The Official Handbook of the Marvel Universe*, edited by Mark Gruenwald.

SECRETS of SPIDER-MAN'S WEB

FORTUNATELY FOR PETER PARKER (AND THE WORLD AT LARGE), THE AMAZING TEEN-AGER IS A BRILLIANT SCIENCE STUDENT! HE HAS DEVOTED LONG HOURS OF STUDY TO LEARNING EVERYTHING HE CAN ABOUT SPIDERS! ALTHOUGH IT IS NOT A MATTER OF PUBLIC KNOWLEDGE, HE IS PROBABLY THE WORLD'S GREATEST AUTHORITY ON THE SUBJECT OF WEBS AND THEIR CREATION...

HIS WEB-MAKING ABILITY IS ONE OF HIS MOST CLOSELY-GUARDED SECRETS! BUT WE CAN TELL YOU THIS... HE MAKES HIS OWN WEB FLUID UNDER THE MOST EXACTING CONDITIONS IN THE LAB, STORING IT IN SMALL, COMPACT CYLINDERS LIKE MINIATURE TOOTHPASTE TUBES!

AS ANY SPIDER-MAN READER KNOWS, SPIDEY'S WEB-SHOOTER IS WORN AT HIS WRIST, AND ACTIVATED BY THE SLIGHTEST TOUCH OF HIS FINGER UPON THE SUPER-SENSITIVE ELECTRODE LOCATED ON THE PALM OF HIS HAND!

INASMUCH AS HIS WEBBING IS HIS MOST POTENT WEAPON, THE MASKED ADVENTURER ALWAYS CARRIES SPARE WEB-FLUID CAPSULES CLIPPED ONTO HIS INGENIOUSLY DESIGNED UTILITY BELT!

BY ADJUSTING THE NOZZLE OF HIS WEB-SHOOTER IN ONE EASY MOTION, SPIDEY CAN EJECT HIS WEB FLUID IN ANY ONE OF THREE DIFFERENT WAYS...

1. AS A THIN, INCREDIBLY STRONG LINE...

2. AS A FINE, QUICK-SPREADING SPRAY...

3. OR AS A THICK, TREMENDOUSLY ADHESIVE LIQUID...!

gate father, then Jameson is the "bad" father in the series, railing at him, refusing to acknowledge the value of his services, but every so often treating him with an awkward parental concern and care. Joe "Robbie" Robertson, Jameson's sensible second-in-command, has proved at times a more sympathetic father figure to Parker; introduced in the mid-1960s, he was one of Marvel's first major African-American characters. Another surrogate father was the wise police captain George Stacy, father of Parker's first true love, Gwen Stacy.

Parker's substitute mother in the series, of course, was his Aunt May. Originally portrayed by Lee and Ditko as an extremely frail and naive old woman who seemed virtually unaware of the world beyond her front steps, May's characterization deepened in recent years. Writer J. M. DeMatteis, especially, portrayed May as a woman of quiet fortitude, bearing the losses in her life with a courage even Peter himself envied.

Among Peter's friends the most notable is Flash Thompson, who in high school hero-worshiped Spider-Man but mercilessly bullied Parker. Over the years Thompson matured and formed a strong bond with Parker. Peter's closest friend, however, was his rich young college roommate Harry Osborn, who became his nemesis. Despite the surface realism of the portrayal of Peter Parker's "normal" friends, each of them has been touched by the world of his costumed identity and in many cases even become part of that world. Unfortunately, there's something ludicrous about what has happened to many of

them over the years. Peter's college biology teacher and his first girlfriend's husband, for example, both became super-criminals: the Jackal and the original Hobgoblin. More often, however, there is a dream logic behind the evolution of the series' supporting characters, as the crises in their lives bring them closer to Peter, as with Flash, or drive them into enmity, as with Harry.

Harry Osborn seemed blessed in ways that Peter was not: he had wealth and a still-living parent. But while Peter grew up amid the love of his uncle and aunt, Harry was neglected by his stern, distant

Amazing Spider-Man #12 (1964) Script: Stan Lee / Art: Steve Ditko

Bennett Brant was no friend of Peter's, but he was the brother of Peter's first girlfriend, Betty. When Bennett became involved with Doctor Octopus, Spider-Man was unable to prevent his death. Betty blamed Spider-Man, and Peter felt that now he could never tell her they were one and the same. Once again, the responsibilities of his life as Spider-Man had ruined Peter's hopes for happiness.

father, the ruthless businessman Norman Osborn, who was secretly the garishly dressed criminal mastermind the Green Goblin. Over time the Goblin and Spider-Man discovered each other's true identities, but during their ensuing struggle the Goblin suffered amnesia; for years afterward, though, Osborn would temporarily revert to his Goblin persona and threaten to take revenge on Spider-Man by killing his loved ones.

Meanwhile, jilted by his girlfriend, Mary Jane Watson, Harry turned to drugs in a story line by Stan Lee and artist Gil Kane, which Lee had published without Comics Code Authority approval in order to take a stand against drug abuse, thereby forcing the Code to alter its position. (The Comics Code Authority, or Code, was formed in the 1950s as a response to congressional investigations of comics.) Finally, the Goblin was accidentally killed in conflict with Spider-Man. Driven mad by grief and chemicals, Harry became the new Goblin, as if the mask of the Green Goblin carried a power that

lived on independently of its hosts. Eventually, Harry seemed cured; he married and fathered a child, but the madness overtook him once more. Recently, in a moving story by writer J. M. DeMatteis and Sal Buscema, he too died, asking Peter's forgiveness. And yet plans he had set in motion as the Goblin still haunt Spider-Man even today.

Though Peter Parker started out as the archetypal bookworm who could not get a date, there have been a number of women in his life. The first was Betty Brant, Jonah Jameson's secretary, who, although this has never been established in the stories, must have been somewhat older than high-school student Peter. They became devoted friends but later grew apart after Betty turned against Spider-Man when he was unable to save her gangster brother Bennett from being killed. Though somewhat dowdy and fragile in spirit in the early stories, Betty has changed over the decades along with pop culture's portrayals of women, and today she is a spirited, aggressive *Bugle* reporter.

Several years after his series began, Peter graduated high school and entered college, where he met his first true love, Gwen Stacy. Eventually, Gwen became an idealized 1960s ingenue, perfectly beautiful and without faults; in retrospect, she seems to have been a rather superficial character. But she was the "good girl," in contrast to the other leading lady of the series, party girl Mary Jane Watson. There had been a long-running subplot during the years Steve Ditko drew the series about Aunt May's attempts to set Peter up with her friend

Amazing Spider-Man #42 (1966) Script: Stan Lee / Art: John Romita, Sr.

Anna's niece Mary Jane. Peter continually resisted, certain that any girl his aunt picked for him would be deadly dull. It was not until artist John Romita, Sr., took over the series that we and Peter finally saw Mary Jane's face: "Face it, tiger," she told him, "you just hit the jackpot!" and she was right. Still, attractive and vivacious as Mary Jane was, she was also portrayed as flighty and superficial, and she ended up briefly going out with Harry, only to break his heart. Peter, meanwhile, fell deeply in love with Gwen, only to see her die in a shocking story by writer Gerry Conway and artist Gil Kane, the successors to the Lee-Romita team.

Subsequent to Gwen's death, Conway hinted at a growing affection between Parker and Mary Jane, but as writers came and went over the years this subplot became lost. Other contenders for Peter's affections arose. Upon finally starting his graduate studies Peter became involved with Debra Whitman, a painfully shy woman who reminded him of himself in his high-school days; their relationship inevitably faded. The opposite extreme was personified by Felicia Hardy, the Black Cat, a costumed thief created by Marv Wolfman and Keith Pollard in the tradition of the many women on the other side of the law who tempt the heroes of popular fiction. Spider-Man was indeed attracted to this woman from his own world of costumed adventurers, but, ironically, she found herself unable to relate to him as the "normal" Peter Parker.

The woman whom Peter Parker finally married proved to be, surprisingly, Mary Jane Watson. The

Amazing Spider-Man #47 (1967) Script: Stan Lee / Art: John Romita, Sr.

JOHN ROMITA, SR.

Peter Parker's romantic life only really took off when John Romita, Sr., replaced Steve Ditko as artist on the series. Romita had considerable background in drawing romance comics, and soon Peter grew more handsome and his girlfriends prettier. Lee, who was not immune to Romita's influence, made his shy hero less repressed, even having him ride a motorcycle; later Spider-Man writer Roger Stern would explain that Parker had simply been a "late-bloomer."

After months of avoiding the blind date Aunt May was trying to arrange for him, Peter finally came face to face with Mary Jane Watson, the girl he would marry, in issue #42, Romita's fourth on the series (top). In his hands, the brunette Watson and the blond Gwen Stacy became Spider-Man's answer to Archie's Betty and Veronica (above).

THE DEATH OF GWEN STACY

In a shocking story line of 1973, the original Green Goblin captured Gwen Stacy and pushed her off the top of a bridge. Spider-Man succeeded in catching her, but the shock of impact broke her neck (left). Three years before, Spider-Man had pledged his honor to her dying father that he would keep Gwen from harm. It was immediately after Gwen's death that Spider-Man had his final confrontation with the Goblin, who accidentally impaled himself on one of his own weapons.

In his *Marvels* series writer Kurt Busiek reprised Gwen's death, marking it as a turning point in Marvel's history, a loss of innocence for the Marvel Universe: from then on happy endings for leading characters were no longer assured, and the Marvel Universe began turning into a grimmer, darker place (below). It should be noted that while the script described the site of Gwen's demise as the George Washington Bridge, the art depicted the Brooklyn Bridge, and there is still no agreement as to where it actually took place.

◄ *Amazing Spider-Man* #121 (1973) Script: Gerry Conway / Pencils: Gil Kane / Inks: John Romita, Sr.

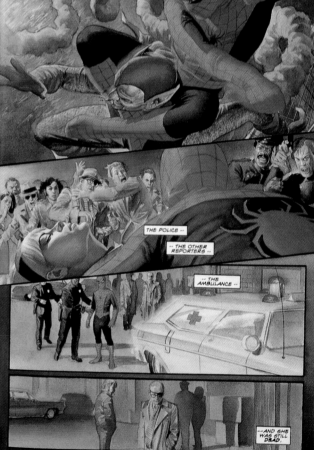

Marvels #4 (1994) Script: Kurt Busiek / Art: Alex Ross

turning point came when the creative team of Tom DeFalco and Ron Frenz had Mary Jane reveal to him that she knew he was Spider-Man; Gerry Conway, in the graphic novel *Parallel Lives*, had her discover his double identity on the very night he apprehended Uncle Ben's killer. DeFalco and Frenz proceeded to fill in Mary Jane's past, portraying her as a deeply unhappy woman who had fled her dysfunctional family in Pittsburgh and who hid her sorrows beneath the happy-go-lucky facade familiar from the *Spider-Man* comics of previous years. Peter and Mary Jane decided to become confidants rather than lovers, but they were resisting the inevitable. On Stan Lee's decision they soon married, a step that was controversial at the time.

Like most characters in comics, Marvel's remain young as the decades pass, but part of the Marvel revolution was that their lives do progress. Peter Parker moved from high school to college while Lee wrote the series and later made the transition to graduate school; he has matured over the years, as have his friends. Marriage was the obvious next step, and despite the expectations of some, the marriage has endured for years.

A DARKER VISION

There are many cases of long-running super heroes whose glory days lie in the past; succeeding writers and artists were never able to match or surpass the greatness of the original creators' stories. That has not proved true of Spider-Man. The classic stories of Stan Lee and Steve Ditko were succeeded by Lee's

Amazing Spider-Man Annual #21 (1987) Plot: Jim Shooter / Script: David Michelinie / Pencils: Paul Ryan / Inks: Vince Colletta

MARRIAGE

Although super hero series of the past almost always featured a love interest, the traditional attitude of super heroes from Superman on down—or, indeed, many heroes in other works of popular fiction—was to exclude the girlfriend, to refuse to confide in her or commit to a relationship. In many of today's comics, there is no love interest for the hero at all, as if a life of endless violence is enough for these characters; many of them are clearly emotional cripples and seem proud of it. In marrying, and thereby affirming their love for each other, Peter and Mary Jane were finally acknowledging and beginning to cope with their emotions. Thus the series came full circle, with Peter and Mary Jane establishing a new center of moral and emotional stability in their unstable world, which took the place of the marriage of the elderly Ben and May that had been destroyed in the very first *Spider-Man* story. Among the wedding guests pictured here (clockwise from upper left) are Betty Brant, *Daily Bugle* editor Joe Robertson, Harry Osborn (who became the second Green Goblin), Flash Thompson (now Peter's friend), Aunt May, J. Jonah Jameson, and Mary Jane's Aunt Anna.

Amazing Spider-Man #257
(1984) Script: Tom DeFalco/
Pencils: Ron Frenz / Inks:
Josef Rubinstein

It appears that the people
who most love Peter Parker
recognize that he and
Spider-Man are two sides
of the same person. Here
Mary Jane reveals she
knows the truth. (Artist Ron
Frenz symbolically portrays
Peter as half-garbed in the
black costume he wore at
the time of this story.)

Amazing Spider-Man #314
(1989) Art: Todd McFarlane

A Marvel Christmas: Peter and
Mary Jane are evicted from their
apartment, and the desperate
hero searches his concience
about asking Aunt May if they
can move back into her house.

collaboration with John Romita, Sr., which pro-
duced such unforgettable tales as the unmasking of
the original Green Goblin and the introduction of the
Kingpin. Among the many other contributors of
memorable additions to the *Spider-Man* mythos
have been writers Gerry Conway, Len Wein, Marv
Wolfman, Roger Stern, Tom DeFalco, David
Michelinie, and J. M. DeMatteis; artists Gil Kane,
Ross Andru, John Romita, Jr., Sal Buscema, Ron
Frenz, and Mark Bagley; and editors Jim Salicrup
and Danny Fingeroth. Special note should be made
of the work of Todd McFarlane, who began drawing
Spider-Man in the late 1980s and succeeded in
recapturing much of the feel that Steve Ditko originally gave Spider-Man and his world, but in his own
innovative and very contemporary style. In their
hands, Spider-Man has retained his immense popularity without changing all that much. The changes
that have occurred, besides those in art styles over
the decades, have been in the world around him. As
the decades have passed, Spider-Man's world, always
one in which innocent old men like Uncle Ben can
be brutally killed, has grown ever darker and more
threatening, reflecting not only the public perception
of the spread of crime but also the outrage against it.

It is not surprising, then, that the most notice-
able change over the years has not been the growing violence by the villains, but that of the other
vigilante "heroes" introduced into Spider-Man's world.
It was as early as 1975 that Gerry Conway and
Ross Andru introduced the Punisher (see chapter
seven) in an issue of *Amazing Spider-Man*. Other

TODD McFARLANE

Spider-Man #1 (1990) Script and art: Todd McFarlane

No artist since Steve Ditko and John Romita, Sr., had so revolutionized the look of the *Spider-Man* series as Todd McFarlane. He greatly widened the eyes of Spider-Man's mask and twisted him into oddly contorted poses so that once again Spider-Man moved like no other Marvel character. He also gave great power to action scenes, an eerily Gothic atmosphere to night scenes, a 1990s sexiness to Mary Jane (far left), an appealing cartoonishness to Parker, Jameson, and other characters, and an element of caricature to the more grotesque villains in Spider-Man's rogues' gallery (such as Stan Lee and Ditko's creation, the Scorpion, below). Together with writer David Michelinie he created Spider-Man's most celebrated recent adversary, Venom.

Amazing Spider-Man #319 (1989) Script: David Michelinie / Art: Todd McFarlane

Amazing Spider-Man #309 (1988) Script: David Michelinie / Art: Todd McFarlane

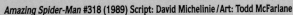

Amazing Spider-Man #318 (1989) Script: David Michelinie / Art: Todd McFarlane

Amazing Spider-Man #316 (1989) Art: Todd McFarlane

When the alien costume bonds with Eddie Brock, Venom develops sharp fangs and his tongue lengthens and hangs lasciviously from his mouth; he longs to kill his opponents, to dismember them, and even to devour their remains (although, thankfully, as yet, he has yet to do more than fantasize about cannibalism).

himself reluctantly allied with both against greater threats, he is determined to bring them to justice. Indeed, Spider-Man increasingly defines himself in the nineties by his stand against such new "heroes." A vigilante himself, Spider-Man refuses to kill and refuses to allow anyone else to act as executioner outside the law.

No characters more incarnate the darkening of the Marvel Universe than two of *Spider-Man*'s most popular recent characters: Venom and Carnage. The fact that *Spider-Man* can still create such exciting new characters demonstrates the continuing vitality of the series.

Venom fills a conceptual gap in the series that apparently no one had noticed before. Virtually every super hero eventually encounters a nemesis who is his evil opposite, a being who embodies his own dark potential. Venom is Spider-Man's nightmarish mirror image, reveling in savagery and bloodshed while protesting his obsession with avenging the innocent.

Venom's origin lay in the twelve-part limited

Amazing Spider-Man #258 (1984) Script: Tom DeFalco / Pencils: Ron Frenz / Inks: Josef Rubinstein

Puzzled by the strange properties of the black costume he acquired on another planet during the *Secret Wars* series, Spider-Man consulted Reed Richards, who told him it was actually an alien being. This was the symbiote that would later join with Eddie Brock to become the monstrous Venom.

vigilantes who prove all too willing to kill without remorse have cropped up over the last decade in the *Spider-Man* books. DeFalco and Frenz introduced the cold-blooded female mercenary leader Silver Sable; she, however, works within the law, and Spider-Man has cooperated with her. The same cannot be said of two of writer David Michelinie's co-creations: Solo, a terrorist who kills other terrorists, and Cardiac, a physician who takes it upon himself to execute people he believes guilty of crimes but who were found innocent by the courts. Though Spider-Man has found

series *Marvel Super Heroes Secret Wars*, in which Spider-Man acquired a new costume during an adventure on an alien world: it was basically a black body stocking with a white spider insignia. Despite Marvel's assertions at the time, Spider-Man did not wear this new outfit for very long: it made him look like a far grimmer character than he was, and inevitably he reverted to his brighter, original red-and-blue uniform.

The black costume, however, literally had a life of its own, proving to be an intelligent, shape-changing alien creature that sought to bond with its host, namely Spider-Man. Writer David Michelinie saw to it that the living costume found a new host, reporter Eddie Brock, a man with a grudge against Spider-Man, and the result was Venom.

This character provides a fascinating variation on the traditional comics theme of the mask and the dual identity. Brock's "mask" is his entire costume, and like Iron Man's it grants him superhuman power. It gives him superhuman strength; it absorbs bullets without harm to Brock; and he can mentally manipulate its shape or use it to camouflage himself. We have already seen how characters alter their behavior in their costumed identities: Peter Parker, for example, becomes more uninhibited, more aggressive, and even more playful as Spider-Man. In Venom's case his costume truly has a mind of its own. When he is bonded symbiotically to this alien creature, Eddie Brock's psyche is influenced by its consciousness: he too loses all inhibitions, but he becomes obsessive, ferocious, and literally bloodthirsty.

Obviously, at first Venom was presented as a villain, the most loathsomely vicious that Spider-Man had yet faced. The surprise was that Venom became so popular that readers wanted to see him in his own series. Readers seemed to enjoy vicariously participating in the release of such savagery.

And they got their wish: Venom now appears in several limited series each year. Venom has mellowed somewhat now that he has been cast in the role of protagonist. He has made himself the protector of an underground community of outcasts beneath San Francisco. His friends there insist he not kill anyone, and so far he is complying, although he still engages in brutality short of murder.

With Venom becoming something of a hero, it was deemed necessary to create an even nastier version of the character to take over his previous role of villain. Hence, Brock's onetime cellmate, a serial killer with the unlikely name of Cletus Kasady, bonded to the spawn of Brock's costume and became Carnage, wreaking slaughter on a massive scale. Spider-Man and Venom have reluctantly teamed up to combat him, and he too has become a tremendously popular villain with the contemporary audience, though he shows no sign as yet of modifying his wickedness.

Amazing Spider-Man #363 (1992) Script: David Michelinie/Pencils: Mark Bagley/Inks: Randy Emberlin Spider-Man continually makes jokes; so does Carnage, the creation of writer David Michelinie and artist Mark Bagley. But Carnage's banter, a stream of consciousness bordering on the nonsensical, like that of many other wisecracking comics villains, is a symptom of his madness. Everything is no more than a joke to Carnage, including the value of human life.

Art: Bill Sienkiewicz and RDA
Carnage

SPIDER-MAN TODAY

In the 1990s absolutely nothing in Spider-Man's life seems stable or certain. Peter Parker's parents, intelligence operatives long believed dead, seemingly turned up alive, only to prove to be androids created by his longtime foe the Chameleon, a master of disguise who himself embodies an unstable identity. Shaken by the revelation, Spider-Man nearly suffered a mental breakdown, seeking violent retribution against the Chameleon and edging ever closer to the line that he had never yet crossed, but that so many other "heroes" of the nineties had: his moral code against killing. Shunning his wife, Spider-Man even rebelled against his own identity as Peter Parker, claiming he wanted to escape painful emotions by living his life only as "the spider," only as the mask. At his lowest point he even retreated into a cocoon of his own webbing. The shock of another personal catastrophe, Aunt May's latest and most serious stroke, brought him back from the brink.

May had begun in the series as a partially comedic figure, overprotectively worrying about her nephew's health, oblivious to the fact that he was Spider-Man. She was also a source of pathos: frail and repeatedly hospitalized, she was a continual concern to Peter. But in recent years writers, especially J. M. DeMatteis, had portrayed her as a pillar of wisdom and stoic endurance beneath her seemingly naive and timorous facade. In her final days she and Peter grew even closer, and she revealed that she had recognized his dual identity long ago

New Warriors #1 (1990) Script: Fabian Nicieza / Pencils: Mark Bagley / Inks: Al Williamson

THE NEW WARRIORS

The popularity of teenage characters in the 1980s led to a whole new family of books at Marvel centering on another team of adolescent heroes: the New Warriors, created by Tom DeFalco and Ron Frenz, and brought to their own series by Fabian Nicieza and artist Mark Bagley. Some, like Firestar (originally introduced on the *Spider-Man and His Amazing Friends* television show) and the new Marvel Boy, were mutants. Other founding members included the wealthy African-American vigilante Night Thrasher and Nova and Namorita, both of whom are discussed in other chapters. The New Warriors are a teenage group of friends and allies, something like a gang, if you will, but one devoted to justice. *The New Warriors* is striking for its ability to shift between radically different moods and tones. On the one hand, there is the sheer comedy provided by founding member Speedball, a 1980s creation of Tom DeFalco and Spider-Man's original artist Steve Ditko, a sexually frustrated high-school nerd who acquires super powers that enable him to bounce around unstoppably. On the other hand, the *Warriors* could shift into a far more serious mode, as with the story of the mutant Marvel Boy. A victim of physical abuse by his father during his childhood, the grown Marvel Boy finally lost control of his powers in a fit of rage and unintentionally killed him. Unmasked, put on trial, and found guilty of manslaughter, he did time in prison, where he reevaluated his life. Like Spider-Man, Marvel Boy's motivation for fighting evil came from recognizing his own moral failure, and he set out in *Warriors* and his own limited series to reclaim his life under the new name of Justice. In recent years *The New Warriors* have become part of the Spider-Man "family" of titles, and the Scarlet Spider became a member for a while.

Amazing Spider-Man #400 (1995) Script: J. M. DeMatteis/
Pencils: Mark Bagley/Inks: Larry Mahlstedt

Spider-Man suffered the ultimate identity crisis when he
encountered his double, Ben Reilly, alias the Scarlet Spider.
By the end of 1995 Ben had taken over the role of Spider-
Man and Peter was leading a normal life with Mary Jane: it
was as if one man with a dual life had been split into two,
each following a different path.

and accepted it. May's death soon afterward, in a
sense orphaning Peter once again, was a dramatic
landmark in the series' history.

Peter was left to face a future in which even his
own identity was uncertain. Years before, in 1974,
the Jackal had created a Spider-Man clone who
Spider-Man believed had been killed in an explosion. Shortly before May's death, a double of Peter
Parker arrived in New York City after years living
away in obscurity under the name Ben Reilly, and
he even became a costumed crimefighter, the
Scarlet Spider. Ben was tormented by the idea that
he was not the real Peter Parker, that though he
cared for Aunt May and others in Peter Parker's life,
his emotions were merely carbon copies of the real
Spider-Man's.

The question inevitably arose: was Ben really
the clone? Or could it be that the Spider-Man whose
adventures had been published since 1974, the one
who had married Mary Jane Watson, was the duplicate? By the end of 1995 it appeared that Ben was
the original Peter Parker, and he took over the role
of Spider-Man. Peter came to accept the idea that
he was the clone, and he and Mary Jane moved to
Oregon, where he began a career as a scientist, the
one he would have followed had he never been bitten by the spider; subsequently, he even lost his
powers. Yet the mystery continued: in early 1996
the clone's skeleton was discovered, again throwing
the identities of Ben and Peter into doubt.

Thus over the years Spider-Man's problems have
escalated from everyday matters such as worrying

KRAVEN'S LAST HUNT

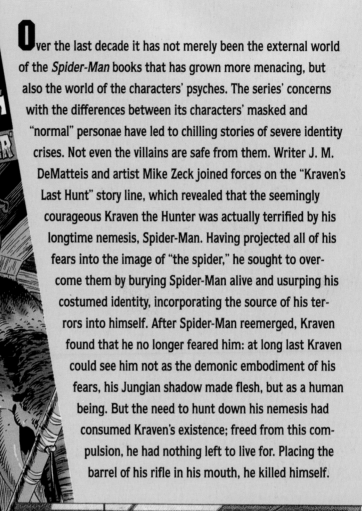

Amazing Spider-Man #294 (1987) Pencils: Mike Zeck / Inks: Bob McLeod

Over the last decade it has not merely been the external world of the *Spider-Man* books that has grown more menacing, but also the world of the characters' psyches. The series' concerns with the differences between its characters' masked and "normal" personae have led to chilling stories of severe identity crises. Not even the villains are safe from them. Writer J. M. DeMatteis and artist Mike Zeck joined forces on the "Kraven's Last Hunt" story line, which revealed that the seemingly courageous Kraven the Hunter was actually terrified by his longtime nemesis, Spider-Man. Having projected all of his fears into the image of "the spider," he sought to overcome them by burying Spider-Man alive and usurping his costumed identity, incorporating the source of his terrors into himself. After Spider-Man reemerged, Kraven found that he no longer feared him: at long last Kraven could see him not as the demonic embodiment of his fears, his Jungian shadow made flesh, but as a human being. But the need to hunt down his nemesis had consumed Kraven's existence; freed from this compulsion, he had nothing left to live for. Placing the barrel of his rifle in his mouth, he killed himself.

Web of Spider-Man #32 (1987) Script: J. M. DeMatteis / Pencils: Mike Zeck / Inks: Bob McLeod

Amazing Spider-Man #400 (1995) Script: J. M. DeMatteis / Pencils: Mark Bagley / Inks: Larry Mahlstedt

Spider-Man's many victories over his criminal opponents have not spared him from suffering personal tragedies. He has been hit particularly hard by the deaths of people he regarded as father and mother figures, beginning with the murder of his Uncle Ben. Among these was Captain George Stacy, the father of Peter's beloved Gwen, and, years later, Aunt May.

One of Stan Lee's most touching scenes came when Captain Stacy was killed by falling debris while trying to save a bystander during a battle between Spider-Man and Doctor Octopus. Dying in Spider-Man's arms, Stacy revealed he knew Spider-Man was really Parker and asked that he take care of his daughter. In keeping with his characteristic bad luck, Spider-Man found himself unjustly blamed for the death of the man he deeply mourned.

The dying Aunt May, the woman who raised Peter after his parents' deaths, likewise revealed, in a dramatic scene on the observation deck of the Empire State Building, that she had seen through to Peter's other, hidden identity (right). A few days later, she finally passed away as Peter recited lines from the story she often told him when he was a child, J. M. Barrie's *Peter Pan* (left).

Amazing Spider-Man #400 (1995) Script: J. M. DeMatteis / Pencils: Mark Bagley / Inks: Larry Mahlstedt

about his next paycheck to the fundamental dilemma of wondering who and what he really is. And still he persists in heroically struggling onward with his life. Throughout the evolution of the series, Spider-Man has remained in essence the same as he was back in 1962. This is the true test of a classic character. And he is more popular now than ever.

In 1962 it probably would have seemed impossible that anyone could have created a new super hero who could rival the popularity and widespread recognition of Superman or Batman. But Spider-Man has done it. He may have been a little guy for whom nothing ever seemed to go right, but he kept on struggling until he won out. But when you think about it, that's the essence of a classic American hero. And that's exactly what Spider-Man has become.

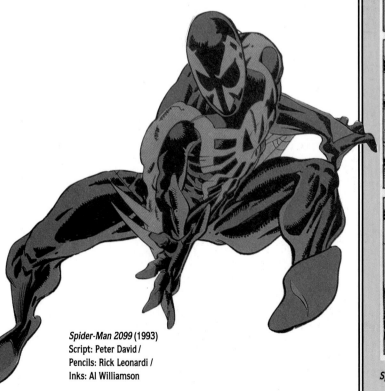

Spider-Man 2099 (1993)
Script: Peter David /
Pencils: Rick Leonardi /
Inks: Al Williamson

SPIDER-MAN 2099

The flagship character of the 2099 line, not surprisingly, is the Spider-Man of 2099, created by writer Peter David and artist Rick Leonardi. In devising futuristic counterparts of Marvel's heroes, the creative teams needed to determine what is archetypal about the original character. They were called upon to vary and update the character while preserving and revitalizing his essence. Unlike Peter Parker, the Spider-Man of 2099, genetic engineer Miguel O'Hara (whose very name illustrates the multicultural nature of 2099's United States), has a brother and a living mother and father: in fact, he has discovered that his real father is his boss, the corrupt Alchemax official Tyler Stone, the man he most detests. Like his predecessor O'Hara has spiderlike abilities, but to a more grotesque degree: the Spider-Man of 2099 fires webbing not from artificial wrist devices, but from spinerettes in his forearms. He sticks to walls using talons on his hands and feet, and he even has retractable fangs with their own venom. At his conceptual heart, though, Spider-Man 2099 is very much the spiritual heir of his twentieth-century namesake: he is a wisecracking, acrobatic free spirit considered an outlaw in a society of enforced conformity, still idealistic despite his surface anger and cynicism, and willing to risk his life to aid the innocent. And as the sequence below demonstrates, he is just as beset by life's continual indignities as the original Spider-Man was.

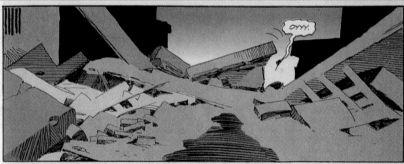

Spider-Man 2099 #16 (1994) Script: Peter David / Pencils: Rick Leonardi / Inks: Al Williamson

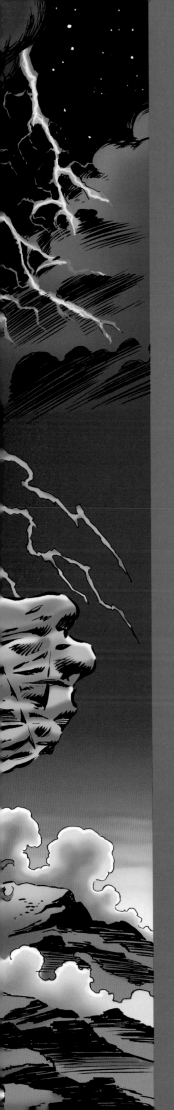

AVENGERS ASSEMBLE!

The story has it that the Fantastic Four was created as a response to National Periodical Publications' best-selling super hero team, the Justice League of America (JLA). In fact the Fantastic Four—a small grouping of friends and family members, all of them new to the comics reading audience—was very different from the Justice League—a team including all the best-known heroes who already starred in their own series at National Periodical (now known as DC Comics).

The actual Marvel counterpart to the JLA was not the Fantastic Four but the *Avengers*, first appearing in 1963, nearly two years after the Fantastic Four's debut. The reason for the delay was obvious: Marvel could hardly team up its best-selling super heroes when it was not publishing any at the time. In the short time since the premiere of the Fantastic Four, however, Marvel had launched four series featuring characters who have remained mainstays of the Marvel Universe ever since: the Incredible Hulk, the Mighty Thor, Iron Man, and the pairing of Ant-Man and the Wasp. These were the heroes who became the founders of the mighty Avengers. (Spider-Man and Doctor Strange had also begun their own series over this two-year period but neither became members. In part this is because the two operated as loners, but it has also been observed that the original Avengers all starred in series drawn by Jack Kirby, while Spider-Man and Strange were drawn and co-created by Steve Ditko.)

There have been many super hero teams created at Marvel over the subsequent years. Certainly, in the 1960s *The Fantastic Four* outshone *The Avengers* in creative brilliance, and today, of course, *The X-Men* far outsells *The Avengers*.

Still, *The Avengers* has always remained one of Marvel's flagship magazines, a comics series to which Marvel's most stellar characters inevitably gravitate either as members or as guest stars.

While the X-Men operate largely in secret due to public mistrust, fighting internecine wars against other mutants, the Avengers are publicly acclaimed as Earth's greatest heroes: a small army that takes on the greatest menaces in the Marvel Universe, spawning epic adventures and intergalactic wars. *The Avengers*, at its best, presents super hero action on the grandest of scales.

THE A TEAM

From Lee and Kirby's perspective, the main problem with the JLA was that the members did not have distinctly different personalities: each was portrayed as an incorruptible, flawless ideal of humanity, and when they all got together, they all spoke and acted in the same manner, as if they were the same person in different bodies. The Avengers would be different.

For one thing, the team originated as the result of a villain's scheme. The greatest enemy of Thor (the Norse god who now fought crime on Earth) was his foster brother Loki, god of mischief and evil. The story began when Loki used his magic to frame the Hulk for nearly causing a train wreck, intending that Thor would learn the news and decide to stop this new menace: as the Hulk's vast superhuman strength rivaled Thor's own, Loki

hoped that the monster would finally defeat his hated brother.

What Loki had not considered was that three new super heroes, Iron Man and the team of Ant-Man and the Wasp, would also hear about the train wreck. Before long the four encountered each other, and while Iron Man, Ant-Man, and the Wasp hunted down the Hulk, Thor went after and captured the real culprit, Loki. Just as everyone was about to go his separate way, Ant-Man suggested they band together to become a virtually unbeatable team. The Wasp even suggested a name: the Avengers. Iron Man and Thor quickly agreed; the Hulk also complied, but for a different reason: by linking himself with these acclaimed heroes in spite of their mutual mistrust, perhaps he would no longer be hunted as an outlaw.

The team put together in the celebratory finale of *Avengers* #1 did not even last a single issue. In the next one, an alien shapeshifter called the Space Phantom decided to play upon the other Avengers' lingering revulsion toward the Hulk. Using his powers to impersonate the Hulk, the Phantom set the other Avengers against him. Issue #2 ended with the Hulk returning to his life as a fugitive, and in the next issue he teamed up with the Sub-Mariner to destroy the Avengers. They failed, and the Sub-Mariner swam off toward the Arctic with the Avengers in pursuit by submarine.

By issue #4, *Avengers* had reached an unexpected turning point, the most important in the series' history before or since. Emerging from the

Avengers #1 (1963) Script: Stan Lee / Pencils: Jack Kirby / Inks: Dick Ayers

AVENGERS ORIGIN

It took four issues for the Avengers to achieve a stable line-up of members. Ant-Man and the Wasp suggested the formation of the team (above). However, wounded by the others' distrust, the Hulk quit the Avengers only an issue later (right). In issue #4 the team rescued Captain America, wearing a tattered soldier's uniform over his costume, in the North Atlantic (below). Frozen in ice and literally worshiped as a god by Eskimos, this mythic hero had been figuratively held prisoner by death and winter until the ice melted and he returned to life as the Avengers' newest member.

Avengers #2 (1963) Script: Stan Lee / Pencils: Jack Kirby / Inks: Paul Reinman

Avengers #4 (1964) Script: Stan Lee / Pencils: Jack Kirby / Inks: Paul Reinman

99

Captain America Comics #1 (1941) Story and art: Joe Simon and Jack Kirby

CAPTAIN AMERICA ORIGIN

"Professor Reinstein" used his "super-soldier serum" to transform the frail, sickly Steve Rogers into a model of physical perfection. But unlike the Aryan superman of Nazi propaganda, this "super-soldier" would serve free peoples, not dominate them. Rogers' new build is impressive and looks realistic. By the 1990s some artists would endow heroes with exaggerated musculature that could not exist in reality.

Arctic waters, the Sub-Mariner came across a small band of Eskimos worshiping a human figure encased in a block of ice. Seized with rage at what he regarded as primitive idolatry, the Sub-Mariner drove off the Eskimos and hurled their frozen idol into the water.

The ice block drifted south, slowly melting, and was finally found and taken aboard by the Avengers, who recognized the man within. It was Captain America, Marvel's greatest and most famous hero of the 1940s, who had been drifting in suspended animation since the end of World War II.

CAPTAIN AMERICA

The national icon, Captain America, was created in 1941 by the team of Jack Kirby and his original partner, Joe Simon, as a response to the outbreak of war in Europe: they clearly wanted the Nazis stopped as much as did their character, Steve Rogers, a frail young man who desperately wanted to join the army. Rogers was clearly and incurably 4-F, but a visiting general overheard him pleading to be allowed to enlist and was impressed by his idealism and courage. The general spirited Rogers away to a safe house in Washington, D.C., where a scientist had him drink "super-soldier serum" and exposed him to "vita-rays," transforming

Tales of Suspense #66 (1965)
Script: Stan Lee / Pencils: Jack
Kirby / Inks: Chic Stone

him into a model of physical perfection. Rogers was to be the first of an army of "super-soldiers" who would win the battle for democracy, but a German "mole" in the lab succeeded in assassinating the scientist and the secret of his serum died with its creator. In the costumed identity of Captain America and equipped with a shield that in the 1960s was revealed to have been made from a virtually indestructible alloy that was created by accident and could never be duplicated, Rogers would be America's only "super-soldier."

Captain America had one of the most unlikely "covers" for a super hero's secret identity in comics: he was a private in the U.S. Army who, in effect, had to go AWOL (absent without leave) every time he went on a mission. His superior, Sgt. Mike Duffy, thought of Rogers as a lazy incompetent and could never understand why the highers-up kept excusing Rogers's repeated disappearances. In the tradition of the highly popular Batman and Robin team, Cap also acquired a kid sidekick, Bucky Barnes, the "mascot" of his army outfit.

The original Simon and Kirby Captain America stories were set in the United States, with Cap and Bucky battling a wide variety of adversaries, some working for the Axis powers and others not. By far the greatest of these was the Nazi saboteur known as the Red Skull. His blood-red skull mask made the Skull not only a powerful visual icon for the Nazis' evil but a living embodiment of death. Simon and Kirby even gave the Red Skull a musical trademark (despite the fact that it could not be heard by comics

Captain America #255 (1981) Script: Roger Stern / Art: John Byrne
(reproduced directly from Byrne's pencils)

REAL AND FICTIONAL ICONS

Adolf Hitler supervised the conversion of an obscure German worker, filled with hatred for humanity, into the foremost symbol of the Nazi terror: the Red Skull (top). In response, the American government created its own living symbol of patriotism, Captain America. Here, America's wartime president, Franklin Delano Roosevelt, presents the captain with his own indestructible shield (above). Though Roosevelt and Hitler are historical figures and Captain America and the Red Skull merely fictional, all have achieved mythic stature as representatives of the opposing systems of democracy and fascism.

Avengers #4 (1963) Script: Stan Lee / Pencils: Jack Kirby / Inks: Paul Reinman

The most traumatic moment of Captain America's life was when he witnessed the death of his partner and surrogate son, Bucky Barnes, at the hands of the Nazi operative Baron Heinrich Zemo.

Avengers #16 (1965) Script: Stan Lee / Breakdowns: Jack Kirby / Finished art: Dick Ayers

Lee and Kirby surely surprised their readership when they wrote all the founding members out of the Avengers and replaced them with "Cap's Kooky Quartet," including three former villains: Hawkeye, Quicksilver, and the Scarlet Witch. The lineup of the Avengers has gone through considerable changes at frequent intervals ever since.

readers): Chopin's "Funeral March," naturally.

Simon and Kirby left after the first year of *Captain America Comics* to go to DC, but Cap went on without them under other hands, including that of Stan Lee, whose first published work for Marvel in the 1940s was a *Captain America* text story. One of the most popular heroes of the "Golden Age of Comics," Cap was also one of the stars of Marvel's first super hero team, the All-Winners Squad, which also included the original Human Torch and the Sub-Mariner and debuted shortly before super hero comics fell from popularity at the end of the decade and Cap and virtually every other costumed character vanished from print.

When Lee and Kirby revived Captain America in *The Avengers* some twenty years later, they picked up the thread of his story toward the end of World War II, ignoring a brief run of Cap and Bucky stories from the early 1950s that had the team fighting Communism. Awakening aboard the Avengers' submarine, Cap shouts for Bucky. It turns out that in the last days of the war in Europe, the duo had tried to stop a Nazi scientist, Baron Heinrich Zemo, from dispatching an experimental plane from England to Berlin. They jumped onto the plane as it took off, but Cap realized too late that Zemo had booby-trapped it. The plane exploded, killing Bucky outright and hurling Cap into the frigid waters of the North Atlantic, where the super-soldier serum in his blood caused him to fall into suspended animation.

Bucky was killed off because Stan Lee had never liked kid sidekicks. What the *Avengers* lacked

Avengers #30 (1966) Script: Stan Lee / Pencils: Don Heck / Inks: Frank Giacoia

as a series, however, was the kind of angst-filled characterization that was serving Marvel so well in other series, and Bucky's death was the key: Captain America became the central figure of *The Avengers*, haunted by the memory of his partner's demise, blaming himself for the tragedy, and feeling lost in a world that had radically changed since World War II. Baron Zemo, hiding in the jungles of South America, learned of the Captain's return and swore to finish him off for certain. He brought together foes of the other Avengers from their individual series as his Masters of Evil, who became the Avengers' most frequent opponents as long as Kirby remained on the series.

That, as it turns out, was not very long. Kirby's last issue (#16), only two years after *The Avengers'* debut, must have come as a shock to readers: Thor was off in Asgard on his own adventures, and the remaining founding members had grown weary of Avengers membership. Ant-Man, who had gained new powers and become Giant-Man, even seemed disillusioned with the life of a super hero and wanted to return to his original career as a scientist. This was certainly unusual behavior for heroes, and so was their next decision: they chose as their replacements three costumed figures who had debuted as super-villains: Hawkeye the Archer, who had menaced Iron Man with the weaponry mounted on his arrows, and Quicksilver and the Scarlet Witch, the

QUICKSILVER AND THE SCARLET WITCH

The mutant twins Quicksilver and the Scarlet Witch first appeared in the *X-Men* as innocent European siblings Pietro and Wanda, whom the sinister mutant Magneto had rescued from a mob terrified by their superhuman powers. Pietro could move at superhuman speed, while Wanda possessed a "hex power" that actually proved to be a psionic ability to manipulate probability, enabling her to make virtually impossible things happen. Obligated to him for their lives, Wanda and Pietro joined Magneto's Brotherhood of Evil Mutants, but ultimately abandoned him and offered their services to the Avengers (above). Ironically, years later they learned that they were actually Magneto's long-lost children.

Quicksilver inherited his father's temper and was infuriated when Wanda married the android Vision; their marriage came to an abrupt end when the Vision was reprogrammed and temporarily lost his capacity for human emotion. Quicksilver wed the Inhuman Crystal and they had a daughter, Luna, but their marriage has been deeply troubled.

Recently, Quicksilver served in the U.S. government's team of mutant operatives, X-Factor, while the Scarlet Witch led Force Works, a team of former West Coast Avengers. Both twins, however, are again active members of the Avengers, as is Crystal. Right, the Scarlet Witch's latest look.

Avengers: The Crossing (1995) Story: Bob Harras and Terry Kavanagh / Art: Mike Deodato, Jr., and studio

Invaders #5 (1976) Pencils: Jack Kirby / Inks: Joe Sinnott
In the 1970s Roy Thomas chronicled new adventures teaming Marvel's Golden Age heroes, Captain America, the original Human Torch, and the Sub-Mariner, during World War II in *The Invaders*.

two reluctant members of the X-Men's foes, the Brotherhood of Evil Mutants. Returning from Zemo's South American stronghold (the villain had paid for Bucky's death with his own at the end of *Avengers* #15), Cap discovered to his surprise that he was now the leader and teacher of a band of young recruits.

Now Stan Lee and Don Heck could devote as much attention to characterization as to action as they delineated the generational clash between Captain America and his younger, hotheaded partners, who slowly developed grudging respect for their mentor.

The Captain also won his own new series by Lee and Kirby in *Tales of Suspense*, which eventually was renamed *Captain America*. Early on they concentrated on recounting Cap's adventures in the European theater of operations during World War II, often adapting and updating stories from the 1940s. Soon they moved the series into the present day, pitting the Golden Age character against futuristic menaces as only Kirby could draw them. The most astonishing of these tales revived the Red Skull, who had come to possess the Cosmic Cube, an object of seemingly unlimited power, capable of transforming thought into reality. Armed only with his courage

and dedication to his ideals, Captain America still managed to face down and defeat evil of unlimited scope and might, as embodied by the Skull.

Subsequent writers continued to expand the legend of Captain America. In the 1970s Roy Thomas, an avid aficionado of the comics of the Golden Age of the 1940s, together with comic-strip artist Frank Robbins, launched *The Invaders*, a series chronicling untold stories of Captain America, Bucky, the Sub-Mariner, the original Human Torch, and Toro against the Axis powers in the early days of World War II.

Next was writer Steve Englehart, who turned his attention to mythologizing the politics of his own time, the first half of the 1970s, the period of the Watergate scandal that drove Richard Nixon from the presidency. A subversive organization from the pages of *Tales to Astonish*, the Secret Empire, set out to destroy the reputation of Captain America, the embodiment of American ideals, as its first step in taking over the federal government. The agents the Empire employed for this task were advertising executive Quentin Harderman (a reference to Nixon's chief of staff, the late H. R. Haldeman, and to his campaign's then new use of advertising agencies) and an organization called the Committee to Restore America's Principles (an allusion to the Nixon campaign's Committee to Re-Elect the President, but with a more vulgar acronym).

This story line ran for months, culminating in Captain America's showdown with the Secret Empire's hooded Number One on the White House

Tales of Suspense #85 (1967) Script: Stan Lee / Pencils: Jack Kirby / Inks: Frank Giacoia

Jack Kirby demonstrates why he is still the comics medium's unsurpassed master of dynamic, cinematic action in this duel between Captain America and the French kickboxer Batroc the Leaper.

lawn. In the confusion Number One fled into the Oval Office; there, unmasked (but with his face hidden from the reader's view), he committed suicide before Cap's horrified eyes. Although it was never explicitly stated, the implication was clear: Number One was the president himself.

The Secret Empire story line segued into a long, introspective period in the series that reflected American disillusionment with the federal government in the aftermath of the Watergate crisis: Cap himself lost faith in his country and for the first time felt obliged to give up his patriotic costumed identity. Instead he created for himself the guise of Nomad, seeking to do good but without reference to any political system. Eventually he resumed his Captain America identity when the Red Skull reemerged on the scene. It appeared that Captain America defined himself less by the standards of the current administration and more by his opposition to everything the Red Skull represented.

For issue #250 the team that created the next memorable period of the series, writer Roger Stern and artist John Byrne, concocted a story in which a third party sought to nominate Captain America to run for president. Although he seriously considered the proposition, Cap finally appeared before the party convention to turn them down. He told them that he could not align himself with any party's specific agenda: he had to represent American ideals that transcended any political platform and had to rely on his own conscience in defining them.

Here is the essence of the super hero myth: it elevates the will and conscience of one man above those of the rest. Steve Rogers is the perfect

Captain America #183 (1974)
Script: Steve Englehart / Pencils:
Frank Robbins / Inks: Frank Giacoia

Captain America briefly assumed a
new costumed identity, Nomad,
during a period when he was disillusioned with the government.

Captain America #250 (1980)
Script: Roger Stern / Pencils: John
Byrne / Inks: Josef Rubinstein

Captain America turns down the
presidential nomination.

American who does not exist in real life, but who
embodies Americans' idealized image of themselves, a Frank Capra hero turned super hero.
Incorruptible, Captain America would never let his
fellow citizens down. Although he lacks superhuman powers, in terms of personality Captain
America is much like DC's Superman: if they suffer
the pains of angst, it is because they are torn
between competing claims on their conscience,
not because they fall victim to temptation from a
darker side. Hence, Cap is less "real" in this sense
than, say, Spider-Man, but his best writers make
him seem real by effectively dramatizing his lonely
struggle to do what he sees as his duty and by
inducing the readers to identify with his pursuit of
moral perfection.

After a period when the series was written by
J. M. DeMatteis and drawn by Mike Zeck, who did
especially admirable work delineating the childhood
and youth of the Red Skull, the writing reins passed
to Mark Gruenwald, who chronicled Captain
America's adventures for ten years, ending in 1995.
Gruenwald sought to create new antagonists for the
Captain that reflected contemporary political and
social trends. Cap had to contend with the anarchist
named Flag-Smasher, the radical feminist Superia,
and the vigilante Blistik, who was out to execute
anyone he felt diminished the quality of life in
today's cities. In keeping with the mood of the
1980s, big business became a particular target. A
group of costumed criminals with identities pat-
terned after snakes, led by the King Cobra, banded

together as the Serpent Society, a satirical variation on a modern corporation, complete with its own health plan. Even old antagonists were reworked along more modern lines: AIM, a subversive organization of scientists, held trade shows at which it sold high-tech munitions to the highest bidder and the Red Skull wore power suits and presided over his many and varied subversive projects from behind a desk, like a chairman of the board.

Gruenwald's Captain America, like Stan Lee's, was out of sync with the times because the limits of acceptable behavior for super heroes had changed. In effect, he is a character from comics' more innocent age of the 1940s in a world where crimefighters like the Punisher kill. He could not even count on the Avengers to follow his moral lead anymore: fully half the team overrode Cap's objections and attempted to destroy the Kree Supreme Intelligence during one of their missions. When he tried to hold an Avengers meeting later to discuss the situation, only three longtime members showed up (ironically including Hawkeye, who once questioned everything Cap did).

At the conclusion of his decade on *Captain America*, Gruenwald ceded the post of writer to Mark Waid, who has made his reputation with his forcefully positive vision of super heroes, in sharp and welcome contrast to the "grim and gritty" characterizations that became a cliché in the comics of the last dozen years. Waid seems well suited for Captain America, who remains after over fifty years the most heroic of all Marvel's super heroes.

Force Works #1 (1994)
Script: Dan Abnett and Andy Lanning /
Pencils: Tom Tenney /
Inks: Rey Garcia

NEED A HAND HERE SCARLET WITCH?

SPIDER-WOMAN AND I WILL BE GLAD TO OBLIGE.

U.S. AGENT

Eventually, Gruenwald had Cap give up his costumed identity yet again. This time it was not because the government he obeyed proved corrupt; it was because he felt he could no longer obey the government. The new federal Committee on Superhuman Activities reasonably decided that since the government created the Captain America costume and identity and supplied Steve Rogers with his shield, they owned the Captain America persona and paraphernalia, as if it were a trademark. Therefore, if Steve Rogers wanted to continue operating as Captain America, he had to do their bidding. Again placing his individual conscience above government policy, Rogers said no and gave up his shield and costume. The commission soon found its replacement: John Walker, who had been trying to supplant Captain America as national hero through stunts for the mass media that he staged in his own costumed guise, that of the Super-Patriot. Walker agreed to become the new Captain America, but he was pathologically withdrawn from most other people, had an inferiority complex, and proved to be ruthless and, worse, unstable. When the Watchdogs, a paramilitary right-wing group, learned who the new Captain America was, they murdered his parents; Walker went berserk and slaughtered many of them; later, he cold-bloodedly killed the two men who had revealed his identity. Steve Rogers, who had by now adopted a new identity and was known simply as the Captain, was outraged by Walker's excesses, but the two were impelled to unite against a greater, mutual foe (no surprise here), the Red Skull. Equally horrified by what Walker had done, the commission was willing to let Rogers have the Captain America identity back on his own terms. Walker himself, realizing the error of his own ways, persuaded Rogers to accept. The lesson here was that Captain America was not the mask, but the man behind the mask; since there was only one Steve Rogers, there could be only one true Captain America. Walker, still basically a troubled loner, repented his past murders and made a new career for himself as the U.S. Agent, first with the West Coast Avengers and then with the team Force Works (above, with the Scarlet Witch and Spider-Woman).

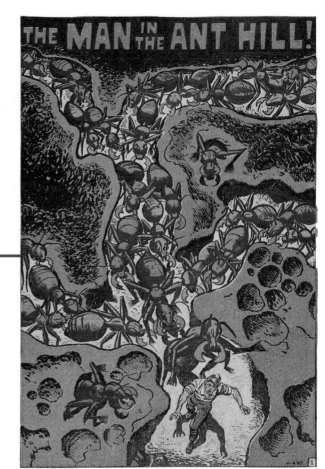

HENRY PYM AND THE WASP

The least well known of the founding members of
the Avengers are Henry Pym, alias the original Ant-
Man, among several other names, and his longtime
partner Janet Van Dyne, the Wasp. Though they
have never achieved true stardom, they have
nonetheless remained active figures in the Marvel
Universe for well over thirty years.

Henry Pym made his debut in "The Man in the
Ant Hill," a characteristic early Marvel science-
fiction tale plotted by Stan Lee, scripted by his
brother Larry Lieber, drawn by Jack Kirby, and published in *Tales to Astonish* #27 in 1962, shortly
after the *Fantastic Four*'s premiere. As the story's
title suggests, Pym was a scientist who discovered a
means of reducing himself to insect size; presumably the story was inspired by the 1957 film *The
Incredible Shrinking Man*. With the success of the
Fantastic Four Lee and Kirby decided to convert Pym
into a costumed super hero as well: Ant-Man, whose
cybernetic helmet enabled him to communicate with
ants, which he could direct as a kind of army.

However, Lee and his collaborators never came
close in his stories to the level of inspiration they
achieved in other series. Ant-Man contended with
various unmemorable Communist agents, the criminal scientist Egghead, whose head was shaped as
his name suggests, and even the Scarlet Beetle, a
mutated bug out to conquer the world. Pym was
Reed Richards without foils for his personality: a
dry, unemotional scientist with little audience appeal.

Lee rescued the character by recasting him as
half of a team. Pym met Janet Van Dyne, the young
daughter of a murdered scientist. When she told
Pym she would do anything to avenge her father's
death, he made her his crimefighting partner. Pym
not only endowed Janet with the power to shrink
but also enabled her to grow wings at insect size,
hence her super hero name of the Wasp.

Although the circumstances of the Wasp's origin
were grim, she quickly evolved into a witty, flirta-
tious character, the opposite of the rather stiff, cerebral Pym. The interplay between their personalities
gave new life to the series, as did Lee and Kirby's
decision to give Pym the ability to grow to gigantic
size as well. They were less creative in renaming
him Giant-Man; after his series in *Tales to Astonish*
ended, Pym returned to the pages of *The Avengers*
under the far better alias of Goliath.

The Wasp fared better as an Avenger than did
Pym; over the years, as attitudes toward women
changed, she evolved from the literally and figuratively flighty ingenue of the 1960s into one of the

KLUK

Marvels #2 (1994) Script: Kurt Busiek / Art: Alex Ross

No other artist has so powerfully captured Giant-Man's Brobdingnagian proportions as Alex Ross did in *Marvels.*

HOW COULD I MIND? YOU'RE OUR GUEST!
AND, AFTER WE AVENGERS INVITED YOU FOUR FOLKS TO COME AND STAY WITH US WHILE YOUR HEAD-QUARTERS IS BEING REBUILT...
WELL, I KNEW IT WOULD BE ASKING TOO MUCH TO EX-PECT THE LEGEN-DARY MISTER FANTASTIC TO JUST SIT ON HIS HANDS ALL THAT TIME.

Fantastic Four #287 (1986) Script and pencils: John Byrne / Inks: Joe Sinnott

Sprightly in personality, the beautiful, winged Wasp resembles a modern-day Tinker Bell.

MARVEL
EARTH'S MIGHTIEST HEROES
AVENGERS
UNIVERSE
ISSUE 394
$1.50
DIRECT EDITION
INTRODUCING THE
WASP?!?

Avengers #394 (1996) Art: Mike Deodato, Jr.

Recently, Marvel decided to give the Wasp a new image: she now has enormous butterfly wings even at her full size, as well as claws, which give her a strangely alien appearance.

Avengers' most effective leaders in the 1980s. Pym, on the other hand, sank slowly into utter failure. The first step in his decline came through what was intended as no more than a clever plot twist. Pym resisted marrying Jan, but exposure to strange chemical fumes altered his personality, and he romanced and married her under the new identity of Yellowjacket. Pym was seemingly cured at the conclusion of the wedding story, but in the1980s *Avengers* writer Jim Shooter made it clear that Pym was still mentally unstable. His marriage to Jan fell apart, and he even struck her in a fit of rage. Pym was court-martialed by the Avengers for irresponsible behavior in combat. Divorced, Pym sank into depression and destitution; the final blow came when his old enemy Egghead framed him for crimes and he was sent to prison. Although Pym was cleared, writer Steve Englehart brought him to the brink of suicide and then saved him: Pym slowly rebuilt his life and today serves again in the Avengers as Giant-Man; at this writing he and Jan have even rekindled their relationship into the closest of friendships.

The classic model of the hero's adventure centers on motifs of death and rebirth. In most super hero stories these motifs are translated into defeat and victory in combat against villains. In Pym's case, readers can trace the gradual collapse of a personality and its redemption. Pym's hero's journey has been more inward, more personal, than most super heroes' adventures, and for that reason it has also been more emotionally affecting.

IRON MAN

Virtually from the beginning of the comics genre there had been wealthy men who became costumed heroes: they had the money to build toys for themselves and they had plenty of time on their hands. And that was the problem: the early heroes were characterized as "millionaire playboys" (though their romantic lives seemed virtually nonexistent) with no real function in business or society. There had also been scientist heroes in super hero stories for decades, who would concoct some strange potion or devise some device that would endow them with superhuman abilities. Science played the same role in these contemporary heroic fantasies as magic had in the heroic adventure tales of centuries past, endowing both protagonist and antagonist with unusual powers.

By 1963, when they created Iron Man, Stan Lee and his collaborators clearly wanted to make the role of the scientist more than a convenient plot device. It was a time of astonishing scientific progress, and ever since the Soviets had launched the first artificial satellite, *Sputnik I*, in 1957, science had become an American obsession. This was the background against which Lee and Ditko could make their teenage hero Spider-Man a science student, and Lee and Kirby could make the dryly cerebral Reed Richards the leader of a super hero team.

There was a place for a glamorous scientist hero, more hands-on engineer than theoretician, a millionaire playboy who was also a captain of indus-try, a man who embodied the early 1960s pride in technological progress and defiance of Communism. That character became Anthony Stark, otherwise known as Iron Man, a brilliant inventor who created an armored battlesuit with which to fight both criminals and foreign agents. The fact that he wore armor made him a modern-day knight, but he was at the same time on the cutting edge of technology, at least as far as his co-creator Stan Lee understood it. Transistors were a recent invention, and the early Iron Man spoke pridefully of the transistors powering his armor.

While inspecting weaponry in the field in Southeast Asia, Stark stepped on a land mine and was captured by rebels led by the petty tyrant Wong Chu. In Wong Chu's camp Stark learned that his wounds were mortal: the shrapnel he had taken would penetrate his heart and kill him in a matter of days. Realizing who Stark was, Wong Chu demanded that he and another captive, the elderly scientist Professor Ho Yinsen, create a new weapon for him before he died. Together, Stark and Yinsen devised an armored battlesuit containing an external pacemaker that would keep Stark's heart beating; moreover, it would amplify Stark's own strength to superhuman levels. The rebels killed Yinsen, but Stark, wearing the armor, wrecked their camp and put them to flight.

Many super hero series owe their fascination, in part, to the clever use of masks and disguises that confer power and preserve the mystique of secrecy, and Iron Man did even more so than others. Stark

IRON MAN IS BORN!

WATCH HIS AWESOME APPROACH! LISTEN TO HIS PONDEROUS FOOTSTEPS AS HE LUMBERS CLOSER... CLOSER... FOR TODAY YOU ARE DESTINED TO ENCOUNTER-- *THE INVINCIBLE IRON MAN!!*

PLOT:
STAN LEE
SCRIPT:
LARRY LIEBER
ART:
DON HECK
LETTERING:
ART SIMEK

Tales of Suspense #39 (1963) Plot: Stan Lee / Script: Larry Lieber / Art: Don Heck

It was Jack Kirby who designed Iron Man's original, bulky armor (above); Iron Man would soon get a more streamlined, formfitting armored body suit and has been wearing variations of this version ever since. Don Heck, however, drew the first *Iron Man* story in *Tales of Suspense* #39 and created the look for Tony Stark, modeling him after Errol Flynn. Stark's public role, though, has similarities to those of the young Howard Hughes, a brilliant multimillionaire inventor with a highly active love life, who remains an enigma to the public at large.

The Vietnam War came early to Marvel: it was clear that Stark's captors (above right) were members of the Vietcong. Since comics characters remain eternally young, Iron Man's origin story posed a problem as the Vietnam War receded into the past. In recent years writer John Byrne slightly modified the origin so that there was no longer any reference to a specific war. In this version, Wong Chu was actually an operative of the Mandarin, a scientific genius who became Iron Man's future archenemy. It has the advantage of explaining why Stark and Professor Yinsen had access to a state-of-the-art science lab (right).

Tales of Suspense #46 (1963)
Plot: Stan Lee / Script: Robert
Bernstein (as "R. Berns") /
Art: Don Heck

The anticommunist theme ran throughout Stan Lee's *Iron Man* stories, which sometimes even portrayed Soviet premier Nikita Khrushchev himself scheming against Stark. (In these panels he meets with a Soviet armored agent, the Crimson Dynamo.)

returned to America, where he fought criminals and subversives in his new secret identity of Iron Man, whom the public believed to be Stark's bodyguard. (This was a first in super hero comics: a hero who was supposedly financed and employed by a private company.) Nor did the world realize that Stark—who appeared to have everything he could wish for—was a man biologically on the brink of death.

Beneath his CEO's power suits and super hero's armor, Stark was a frail, vulnerable man. Though Iron Man could fly through the air, Stan Lee's stories often found him lying prone and nearly helpless on the floor, having depleted his battlesuit's power supplies. And if the battlesuit ran out of power, his heart would stop.

Similarly, though Stark seemed to be a promiscuous playboy without deep feelings for anyone, in fact he (like so many other Stan Lee heroes) was suffering from unrequited love, in this case for his devoted but somewhat plain (until artist Gene Colan glamorized her) secretary Pepper Potts. (Now there is the kind of early 1960s character name no one would dare come up with today.)

During Stan Lee's time on the series, he and artist Don Heck created two of Iron Man's most persistent nemeses. One was the Titanium Man, Iron Man's Soviet counterpart, a giant wearing even more formidable armor than he. The other is one of Marvel's greatest villains, the Mandarin, who has

remained through four decades Iron Man's premier foe; indeed, he was the master villain in the *Iron Man* segments of 1994's *Marvel Action Hour* animated television series.

A Chinese nobleman who refused to accept the Communist revolution, the man who would call himself the Mandarin stumbled across the wreckage of a starship once piloted by beings resembling the dragons of Chinese mythology. (In fact, years later, John Byrne would reveal that Lee and Kirby's infamous dragon Fin Fang Foom from their early monster comics was a member of this race!) In studying the ship, the Mandarin mastered its futuristic technology; he also discovered within it ten rings with amazing powers that he thenceforth wore as his personal weapons. Ever since then the Mandarin has attempted to create a new Chinese empire that would encompass the entire world, even if it means launching a third world war in the process.

Stark's recurring heart problems had already become a cliché by the late 1960s, and Stan Lee's successor as *Iron Man* writer, Archie Goodwin, disposed of the problem by having Stark undergo an artificial heart transplant. But as usual, a strong series concept will persist, and if the form it takes wears out, it will eventually take a new one. The team of David Michelinie and Bob Layton came up with a new physical problem for Stark: the stresses of his dual life drove him into alcoholism. Stark seemingly kicked the habit within one issue, but the theme proved irresistible. Writer Denny O'Neil pro-

HAWKEYE AND THE BLACK WIDOW

Tales of Suspense #52 (1964)
Plot: Stan Lee / Script: N.
Korok / Art: Don Heck

Two of the most familiar Avengers, Hawkeye and the Black Widow (below), began, respectively, as a sideshow archer and as a seductive Soviet spy (above) who contended against Iron Man, first individually and then as a team. Hawkeye equipped himself with an endless array of trick arrows that might release knock-out gas, clamp an opponent's hand to a wall, or even detonate on impact. A consummate athlete, the Black Widow soon adopted a costume and her own weapon, her "widow's bite," that could emit a stunning electrical charge. This seemingly mismatched couple fell in love and eventually abandoned their Soviet masters.

Hawkeye petitioned for membership in the Avengers, but once admitted, he continually challenged the authority of team leader Captain America, whom he mocked as an over-the-hill "Methuselah." As time passed, though, Hawkeye and the Captain became friends and confidants, and, to Hawkeye's astonishment, he eventually found himself leading the Avengers' West Coast branch, which lasted for many years in the Los Angeles area.

The Black Widow defected to the United States and has spent years as a freelance operative for Marvel's intelligence agency, S.H.I.E.L.D. She broke up with Hawkeye and had a long romance and partnership with the costumed vigilante Daredevil (right, in artist Frank Miller's costume). In recent years she has served as the Avengers' leader, and, ironically enough, this former KGB agent has also become one of Captain America's closest allies and friends.

Art: Joe Chiodo

Avengers: The Crossing (1995)
Story: Bob Harras and Terry
Kavanagh / Art: Mike Deodato,
Jr., and studio

113

Tales of Suspense #84 (1966) Script: Stan Lee /
Pencils: Gene Colan / Inks: Frank Giacoia

The self-sacrificing Stark was in continual trouble
with Congress, where Senators misconstrued his
every move. Finally forced to testify before a
Senate committee, Stark suffered a heart attack,
and the fragile condition of his health at last
became public knowledge.

Tales of Suspense #93 (1967) Script: Stan Lee /
Pencils: Gene Colan / Inks: Frank Giacoia

The threat of a massive Soviet arms buildup
during the Cold War seemed embodied by
Iron Man's Soviet counterpart, the gigantic
Titanium Man.

Pencils: Bob Layton /
Inks: Catherine Bollinger

As Tony Stark is an inventor, it
is not surprising that he has
changed Iron Man's armor

vided a more daring and realistic treatment: manipulated in both his love life and business dealings by a new rival, Obadiah Stane, Stark returned to the bottle, only to lose everything—Stane bought out his company and had Stark's assets frozen. The handsome playboy millionaire ended up as a homeless drunk living on Manhattan's streets.

Meanwhile, the role of Iron Man had been taken over by Stark's former right-hand man, James "Rhodey" Rhodes, a character created by Michelinie and Layton. Rhodes had been a soldier who had fought alongside Iron Man in Asia right after he had escaped Wong Chu's camp; Stark subsequently hired Rhodes to be his personal pilot, and he ended up helping Stark and Iron Man in their various adventures. Rhodes continued as Iron Man while Stark hit bottom and slowly climbed back to sobriety and the top of his profession. In keeping with the times, Stark and Rhodes moved out to California to start up their new high-tech firm in Silicon Valley. Eventually, when Stane launched an all-out assault against Stark and his friends, Stark resumed his Iron Man identity and defeated Stane (who had developed his own battlesuit) in combat. Stane committed suicide, and Rhodes returned to his role as Stark's aide. Stark has succeeded ever since in controlling his thirst for alcohol, but it remains a continual, threatening presence in his life.

Even this was not the end of Stark's physical ailments. In a "fatal attraction" scenario devised by Michelinie, Stark was shot and temporarily disabled

▶ *Tales of Suspense* #86 (1967)
Script: Stan Lee / Pencils: Gene Colan / Inks: Frank Giacoia

Iron Man #275 (1991) Script: John Byrne / Pencils: Paul Ryan / Inks: Bob Wiacek

THE MANDARIN

Obviously the Mandarin owes a debt to Sax Rohmer's creation Fu Manchu (who in the 1970s would himself turn up in Marvel's *Master of Kung Fu* series), as well as to Lee and Kirby's 1950s archvillain, the Yellow Claw. Some early depictions of the Mandarin veer uncomfortably close to racial stereotyping (top); his visualizations in the 1980s and 1990s steer clear of this problem (above). The early 1990s *Iron Man* writer Len Kaminski shifted the Mandarin's attention away from science toward mysticism. The opposition between Iron Man and the Mandarin thus becomes one of progressive ideals versus archaic habits of thought.

Panel dialogue (Iron Man #118):

UHHNN. MUST'VE... PASSED OUT! RUSH-ING AIR...*REVIVING* ME! BUT WHERE--?

OH-- --MY-- --GOD!

I WAS RIGHT! SHIELD *IS* AFTER ME! AND THEY'VE *GOT* ME UNLESS I CAN GET INTO MY *ARMOR!*

DAMN! SOMEONE'S GOING TO *PAY* FOR THIS!

THAT WAS A $300 PIERRE CARDIN *JACKET* I JUST TOSSED AWAY!

G-GOT TO HURRY! THAT'S *LAND* BELOW ME NOW, NOT WATER!

THOUGH FROM THIS HEIGHT, I GUESS A *SPLASH* IS THE SAME AS A *SPLAT!*

ONLY THE HELMET TO GO! B-BUT THE GROUND'S GETTING *CLOSE!* GOT TO MAKE THE LAST *CONNECTION,* ACTIVATE THE CIRCUITS--

22

40¢ 128 NOV 02454

MARVEL COMICS GROUP

IRON MAN

DEMON IN A BOTTLE!

Iron Man #128 (1979) Art: Bob Layton
Stark's genius gave him the power of Iron Man, but his body's weaknesses are his worst enemy. Even after he received his heart transplant, he fell victim to the alcoholism he fought repeatedly for years.

Iron Man #118 (1979)
Script: David Michelinie /
Breakdowns: John Byrne /
Finished art: Bob Layton

Time and again the *Iron Man* series dramatizes the contrast between Stark's extreme vulnerability in his everyday identity and his invincibility once garbed in his armor.

by a stalker, Kathy Dare. Attempts to cure him worked for a time, but under writer Len Kaminski, they finally resulted in Stark secretly going into sus-pended animation to stave off his death while work proceeded on a new artificial nervous system for him. The operation proved to be a success, but for months Stark lay bedridden, directing a robotic Iron Man battlesuit by remote control.

For years the Iron Man concept—the 1960s-style playboy who had made a fortune in muni-tions—seemed to some to be dated. In the 1990s,

Kaminski reconceived Stark as an ideal corporate leader, in contrast with the evil businessmen of the 1980s. Stark propounds his enlightened manage-ment theories within the comics, and the stories have emphasized the many characters developed as Stark employees over the decades. Moreover, recent stories have kept technical references up-to-date and even introduced elements of cyberpunk fiction into the series. With the surge of interest in comput-ers, robotics, and biotechnology, Iron Man suddenly seems a more relevant creation than ever.

WAR MACHINE

When Jim Rhodes, an African-American, took over the costumed identity of one of Marvel's flagship characters, it was another comics first. Archie Goodwin had had a black man, Eddie March, substitute as Iron Man for a few issues, and Denny O'Neil and Neal Adams had come up with a black "alternate" Green Lantern at DC. But Rhodes remained Iron Man in Tony Stark's stead for years. Later in the *Iron Man* series, after Stark had resumed his role as Iron Man, Rhodes used one of the Iron Man battlesuits to become the title character of the series, *War Machine* (below). In this guise Rhodes became a one-man army intervening in wars and crises all over the world, whether the U.S. government liked it or not. Eventually, War Machine distanced himself further from Iron Man by acquiring a new battlesuit of extra terrestrial origin (left).

War Machine #22 (1995)
Art: Sergio Cariello

War Machine #1 (1994) Script: Scott Benson and Len Kaminski / Pencils: Gabriel Gecko / Inks: Pam Eklund

THE MIGHTY THOR

Despite their interest in the scientific wonders of the 1960s, Lee and Kirby also recognized that in creating super hero adventure stories, they were dealing with the stuff of myth. With the exception of *Wonder Woman*'s Olympian gods, few attempts had been made to integrate actual myths from past cultures into the comics genre. Lee and Kirby turned to the warrior heroes of Norse mythology and brought the god of thunder, Thor, from Asgard to contemporary New York City. In the process they developed a theme familiar to religions past and present: the god who descends to Earth and lives among mortals in human form.

The new saga began in 1962 in the pages of *Journey into Mystery* #83. While exploring the wilderness on a vacation in Norway, the frail, disabled American physician Don Blake came across an invasion party of aliens, the Stone Men of Saturn. Trapped in a cave, Blake found a hidden door leading to a secret chamber, where a gnarled wooden walking stick lay. In frustration he struck the stick against the cave wall, there was a flash of light, and he found himself transformed into a tall, heavily muscled figure with long, blond hair (which had not yet become fashionable when this story first appeared in 1962), wearing a winged helmet, a cape, and a strange costume. The cane had metamorphosed into an ancient hammer with the inscription "Whoever holds this hammer, if he be worthy, shall possess the power of THOR!"

Overjoyed, Blake realized that he had become the thunder god worshiped by the Vikings. Thus reborn, Thor came forth from the cave and single-handedly drove the Stone Men off the planet.

Blake returned to his medical practice in New York City, but continued to transform into Thor to combat menaces ranging from ordinary gangsters to Communists to super-villains like Mister Hyde and the Cobra. Blake's role as a gentle healer provided an effective contrast with his warrior alter ego, but within three issues it was already becoming clear that the relationship between Blake and Thor was not what it had first appeared to be. Blake was not a quiet human who had by chance gained a god's superhuman powers. It took Lee and Kirby until 1969 finally to make explicit the deeper identity of Blake and Thor. It turned out that, in a grand case of the pot calling the kettle black, Thor's father, Odin, had decided to teach his arrogant son humility by transforming him into a disabled Earthling with no memory of his godly heritage. In recovering his hammer in the cave, Blake was simply regaining his true identity, according to Odin's plan.

As the Thor series evolved, Asgardian characters played increasingly larger roles. Thor, it was discovered, had a deadly enemy, his foster brother Loki, god of evil and mischief, who followed him to Earth to renew their blood feud. Loki is another of Marvel's great villains, the company's most successful recasting of the trickster figure of world mythology, constantly assuming new guises and concocting new schemes to manipulate his ene-

mies. The clashes between Thor and Loki represent a thoroughly satisfying treatment of a favorite device of Lee and Kirby's, the motif of sibling rivalry, with the two brothers embodying the good and evil sides of the human personality. Intensely jealous of his brother's nobility, Loki continually schemes to destroy Thor—and once even switched bodies with him—to unseat their kingly father and rule Asgard, home of the gods.

Although Thor is always ultimately loyal to him, his father, Odin, has often plagued him as much as Loki. Odin is the stern father of childhood fears writ large: aged but robust, resembling a bad-tempered Santa Claus, Odin was absolute ruler of Asgard and demanded total obedience, especially from his son and heir. Monstrously egotistical, Odin was forever proclaiming he was "all-wise" and "all-knowing"; worse, since he possessed seemingly unlimited cosmic power, he could not be defied, except by beings of equal power, like his archfoe, the ancient fire demon Surtur. Odin embodied the forces of order prevailing against Loki and the other agents of chaos, who, according to the myths of Ragnarok that Lee and Kirby recounted, would someday destroy Asgard and its gods. This simply made it harder for Thor to disobey Odin's unjust decrees.

After a few years, Lee and Kirby introduced the "Tales of Asgard" series into the back of Thor's own magazine. Here, they adapted the original Norse myths, told new tales of Thor's childhood, and eventually presented lengthy epics of Thor's quests in far-off reaches of the Asgardian continent.

Journey into Mystery #83 (1962) Plot: Stan Lee / Script: Larry Lieber / Pencils: Jack Kirby / Inks: Dick Ayers

THOR ORIGIN

Trapped in a cave and facing death, the disabled Don Blake was reborn as the mighty Thor, as his new-found gnarled and crooked cane underwent a simultaneous Freudian transformation into a powerful hammer.

By the time *Journey into Mystery* was renamed *Thor* in 1966 Lee and Kirby had brought the series to its first creative peak. As in their simultaneous work on *The Fantastic Four*, Lee and Kirby were crafting comics masterworks, combining spectacular graphics with a ceaseless flow of amazing concepts. While Thor continued to face adversaries like the Absorbing Man and Wrecker on Earth, he also led his fellow gods in combat in Asgard. High points include the invasion of Asgard by the troll king Geirrodur, who had forced his captive, the godlike alien Orikal, to do his bidding, and the desperate struggle to prevent the monstrous Mangog, the incarnation of the rage of a race of Asgard's foes, from drawing the mystical Odinsword, an act that would destroy the cosmos. Yet another body of mythology was tapped as Thor met the Greco-Roman demigod Hercules as a rival in combat and thereafter descended into the netherworld of Hades to rescue him from enslavement by the death god Pluto.

The creative team also devised astonishing science-fiction adventures for Thor. The thunder god journeyed into the Black Galaxy to confront Ego the Living Planet, a world with the face of a bearded madman. Atop Mount Wundagore in the Balkans Thor discovered the High Evolutionary, a geneticist who had played both God and King Arthur, evolving humanoid beings from animals and garbing them in armor as his Knights of Wundagore.

When Jack Kirby left *Thor* in 1970 (with Lee also departing the next year) he proved a hard act to follow. In fact, nobody succeeded in matching the glory days of the series until Walter Simonson took over as writer and artist in 1983. Simonson revitalized the look of the series with his handsome, stylized, but powerful graphics. He drew upon an

Thor #493 (1995) Script: Warren Ellis / Art: Mike Deodato, Jr.

The royal family of Asgard: from left to right, Thor, the warrior prince; his "all-powerful" father, Odin, monarch of the realm; and Thor's foster brother and eternal rival, Loki, god of evil. In Marvel's mythology they are the most powerful members of a race of other-dimensional beings whom Earthmen once worshiped as gods.

extensive familiarity with the actual Norse myths without ever becoming tiresomely didactic. While actually deeply faithful both to the spirit of the original myths and to Lee and Kirby's work, Simonson delighted in taking an outwardly irreverent attitude and was willing to turn any preconceived notion about the familiar characters on its head.

For example, in his very first issue somebody else proved worthy of lifting the magic hammer and gained the power of Thor—a horse-faced alien cyborg with the unlikely name of Beta Ray Bill—leaving the mortal Blake stranded on Earth. Thor eventually regained his hammer and godly powers, while Odin granted Bill his own hammer. But the Blake persona had outlived its usefulness; rather than change into a frail mortal, Thor now masqueraded on Earth as a bespectacled construction worker, a builder rather than a healer. In time, he grew a beard to hide scars inflicted on his face and ended

up looking like an idealized version of his writer/artist.

Simonson placed Thor in some of his most thrilling adventures in the history of his series, blending the contemporary and the legendary in surprising ways. Imagine Thor leading the Einherjar, the spirits of the dead Viking warriors of Valhalla, in an invasion of the realm of Hela the death goddess, the warriors all armed with automatic weaponry from Earth. Or witness Fafnir, an evil monarch whom Odin had turned into a Brobdingnagian dragon centuries ago, rampaging through New York City only to face Thor and an old man who was the last worshiper of the Norse gods; the man died in the great battle that ensued, and Thor gave him a Viking funeral atop Fafnir's flaming corpse.

When Simonson left Thor in the late 1980s, the former Amazing Spider-Man team of scripter Tom DeFalco and Ron Frenz took over. Unexpectedly,

KIRBY'S ASGARD

As the awestricken human feels his senses reeling before the monumental grandeur of what he beholds, there is--there can be--*no thought* of picture-taking--no thought of anything save the...onderment which fills...s soul...

Walk you now at my side, mortal--with slow and measured tread!

Journey into Mystery #123 (1965) Script: Stan Lee / Pencils: Jack Kirby / Inks: Vince Colletta

TALES OF ASGARD, HOME OF THE MIGHTY NORSE GODS

"THE FATEFUL CHANGE!"

My glorious record proves that Volstagg knows not the meaning of *fear!* But why do we race so quickly into a land where *death* lurks everywhere?

Yon city of *Muspelheim* has fallen to the hordes of *Harokin!* We must *liberate* the land while we can!

True, Hogun! Yet, I perceive no signs of *battle!* How did Harokin *achieve* his conquest?

There can be but one answer, Fandral---

...Harokin has seized the enchanted *Warlock's Eye!* Possessing it, *none* can stand against him!

Script: **Stan Lee**
Art: **Jack Kirby**
Inking: **Vince Colletta**
Lettering: **Sam Rosen**
Costumes: **Asgard Haberdashery**

Thor #130 (1966) Script: Stan Lee / Pencils: Jack Kirby / Inks: Vince Colletta

Kirby created a stunningly memorable portrait of the city of Asgard: fantastic palaces atop an island floating through outer space in a dimension adjacent to Earth, connected to it by Bifrost, the Rainbow Bridge from the original myths (above). Lee and Kirby populated this city with an enormous supporting cast, some drawn from the myths and others of their own devising. The most popular were the Warriors Three (left, with Thor), who may have been Lee and Kirby's attempt to work the Three Musketeers into Thor's world. There were (from right to left) Fandral the Dashing, a swashbuckling swordsman in the style of Flynn and Fairbanks; Hogun the Grim, a brooding warrior with a mace who hailed from a kingdom outside Asgard; and finally, the enormously obese Volstagg, who was quite obviously based on Shakespeare's Falstaff. Like his great predecessor, Volstagg was fat, jovial, and boastful, proclaiming himself to be "the Lion of Asgard." In fact, he was often (though not always) a coward, apt to achieve his victories by knocking over his adversaries with his great bulk in trying to escape from them.

Journey into Mystery #121 (1965) Script: Stan Lee / Pencils: Jack Kirby / Inks: Vince Colletta

Loki transformed an ordinary convict into the Absorbing Man, who can absorb the properties of anything he touches. In this panel he grows in size, becoming a modern counter-part to the giants who fought the gods in Norse myths.

▼ *Journey into Mystery* #126 (1966) Script: Stan Lee / Pencils: Jack Kirby / Inks: Vince Colletta

Having brought Norse mythology into the Marvel Universe, Lee and Kirby inevitably introduced the pantheon of Greek and Roman gods as well. They characterized Hercules very differently from the regal Thor: to him, combat was just another of his many pleasures, along with wine, women, and food. But when forced by circumstances, he proved just as noble and valorous as Thor. Rivals early on, Thor and Hercules became staunch allies, and Hercules has frequently served in the Avengers. (Notice how Kirby turns up the intensity of this battle scene through this tightly framed full-page shot.)

Thor #353 (1985) Script and art: Walter Simonson

Walter Simonson successfully recast elements of Norse mythology into a dramatic, contemporary visual style. Here the gigantic fire demon Surtur prepares to lay waste to both Asgard and Earth.

Thor #134 (1966) Script: Stan Lee / Pencils: Jack Kirby / Inks: Vince Colletta

The High Evolutionary was a human genetic engineer who turned animals into "New Men" that he organized into knights of a latter-day Round Table. Eventually he turned his technology on himself and became a being of godlike power.

they decided to return to the implicit concept of the very first story: the normal man who gains the power of a god. Early in their run they introduced Eric Masterson, a divorced architect with a young son, whose life began intersecting with Thor's. When Thor finally slew Loki in combat—need I add that Loki did not stay "dead" forever?—Odin punished him by imprisoning the thunder god's consciousness within Masterson's subconscious (Masterson could still change into the physical form of Thor while retaining his own personality).

Now DeFalco and Frenz could examine what Lee and Kirby had not: how does a person with no experience as a fighter cope with suddenly inheriting the powers and responsibilities of a superhuman crimefighter? The new Thor made mistakes, suffered self-doubt, was called on the carpet by Avengers leader Captain America, often thought of himself as a loser, but nevertheless proved a success. Eventually, of course, the real Thor returned to physical form, but he and Odin were grateful to Masterson and presented him with his own weapon, the mace called Thunderstrike. Using it Masterson could still transform into a godlike being, albeit one who was only half as powerful as Thor. Taking his new *nom de guerre* from his mace, Masterson fought evil in his own *Thunderstrike* comic book until he perished heroically in its final issue.

Thor #367 (1986) Script and pencils: Walter Simonson / Inks: Bob Wiacek

Lee and Kirby told the story of the Norse god Balder, loyal to Thor but perhaps more noble even than he, who became the unwilling object of the love of the Norn Queen Karnilla, a sorceress often allied with Loki. After years of prudishly resisting Karnilla's advances, Balder finally gave in and she proved not to be so reprehensible after all; Balder's quest to rescue her from captivity by the Frost Giants became the subject of Simonson and Sal Buscema's remarkable Balder the Brave limited series.

Recently, writers Warren Ellis and William Messner-Loebs, along with artist Mike Deodato, Jr., have taken Thor in a different direction. They put Thor back on Earth, gave him a new costume (seen in this chapter's frontispiece), and even dropped his use of archaic words like "thee" and "thou." Isolating him from Asgard, Ellis and Messner-Loebs emphasized the paradox of a god of ancient times dwelling among contemporary mortals.

WALTER SIMONSON

Thor #337 (1983) Art: Walter Simonson

Thor #380 (1987) Script and layout: Walter Simonson /
Finished art: Sal Buscema

Thor #343 (1984) Script and art: Walter Simonson

A master entertainer, writer/artist Walter Simonson kept *Thor*'s readers off balance with unexpected plot twists, firing their enthusiasm to see what would happen next. In the very first issue he wrote, for instance, the alien Beta Ray Bill temporarily supplanted Don Blake in the role of Thor (above left). At its best Simonson's *Thor* conjured a sense of epic adventure in modern times. One of the finest examples is the conclusion of the tale of Thor's only worshipper on Earth, when "The Last Viking" died heroically in battle and was conveyed to Valhalla by the Valkyries (above right). Here Odin appears in his guise as the Wanderer, familiar to those who know Richard Wagner's opera *Siegfried*. Simonson left his mark in Thor in another way, giving the thunder god not only a suit of battle armor, but also a beard resembling the artist's own, as seen in an unusual issue consisting wholly of full-page panels depicting a battle in which Thor slew the Midgard Serpent of the Norse myths (left).

Thor #365 (1986) Script and art: Walter Simonson

The most celebrated example of Simonson turning the book temporarily upside down came when he had Loki transform Thor into a tiny frog. For the next few issues Thor, retaining his normal mind, became involved in a war between the frogs and rats of Central Park, and he even won the heart of a young frog princess along the way. Simonson then topped himself by having Thor find his hammer and use it to restore his normal size, powers, and costume—except that now Thor was a god-sized frog! The Frog of Thunder mounted his enchanted chariot and rode to Asgard, where he forced Loki to return him to normal. It was all ludicrous on the surface, but absurdly funny, and yet Simonson also managed to fill it all with real suspense and drama, and the charm of a classic children's fairy tale, even arousing sympathy for the frog princess's heartbreak when Thor left her.

Thor #493 (1995) Script: Warren Ellis / Art: Mike Deodato, Jr.

From Kirby to Simonson to Jackson Guice to Mike Deodato, Jr., Amora the Enchantress has drawn out her artists' skills at evoking feminine seductiveness. Long an enemy of Thor and other Marvel heroes, this Asgardian sorceress at last became Thor's lover in 1995.

Thunderstrike #2 (1993) Script: Tom DeFalco / Pencils: Ron Frenz / Finished art: Al Milgrom

Endowed with powers resembling Thor's, Eric Masterson blended elements of the thunder god's traditional costume with a 1990s streetwise look in his role as Thunderstrike. Surprisingly, Thunderstrike was allowed to die when his series met its end in 1995.

Squadron Supreme #1 (1985) Script: Mark Gruenwald / Pencils: Bob Hall / Inks: John Beatty

SQUADRON SUPREME

In the 1980s writer Mark Gruenwald's *Squadron Supreme* series made them far more than a running joke: taking the concept of super hero vigilantism to its limit, he showed the Squadron becoming the virtual dictators of their world's America, turning it into a utopia but inevitably provoking a revolution against their benevolent tyranny. In the center of this panel stands Squadron leader Hyperion. To the left of him in the far background are the dwarfish genius Tom Thumb and Nuke; in front of them are the light-haired super-speedster the Whizzer, Doctor Spectrum, the dark-haired Power Princess, and the blonde sorceress Arcanna. To the right of Hyperion are the Golden Archer, Lady Lark, Amphibian (with his hand raised), and the winged Blue Eagle. On the far right, out of costume, is the millionaire crimefighter Nighthawk, who led the rebellion that finally overthrew the Squadron.

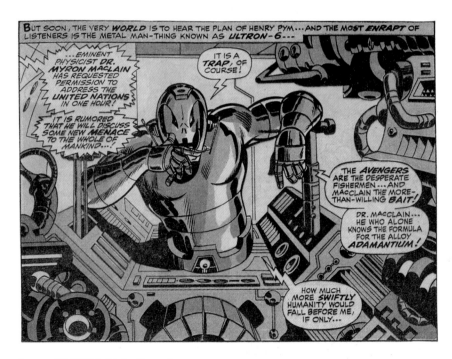

Avengers #68 (1969) Script: Roy Thomas / Pencils: Sal Buscema / Inks: Sam Grainger

One of the Avengers' most formidable foes, the nearly indestructible robot Ultron, longs to destroy his human "father" Henry Pym and wipe out the human race.

THE WAR OF THE GALAXIES

At the end of 1966 Stan Lee turned the writing chores on *The Avengers* over to Roy Thomas, who quickly seized the opportunity to show that he could stage spectacles rivaling those in Lee's other books. Whereas Lee's recent Avengers had a relatively intimate scale, Thomas expanded the size of the team, adding new members such as Hercules, the Black Panther, and a new version of the Black Knight; eventually Lee allowed him to bring Thor and Iron Man back to the ranks as well. Thomas and artist John Buscema created one of the Avengers' greatest adversaries, the virtually indestructible robot Ultron, who was possessed by a Freudian hatred for his inventor, Henry Pym. Thomas and Buscema also dispatched the Avengers to an alternate universe where they first met the Squadron Supreme, a super hero team wittily evoking other.

Within the illustration:

ALL WE KNEW WAS THAT THEY WERE IN ANOTHER GALAXY.

ALL WE KNEW WAS THAT THEY WERE FIGHTING IN A WAR -- A WAR BETWEEN TWO ALIEN RACES. AND IF THEY FAILED, WE WOULD ALL DIE.

THERE WERE NO BULLETINS. NO RADIO REPORTS. HOW COULD THERE HAVE BEEN?

ALL WE COULD DO WAS WAIT. AND HOPE.

Marvels #3 (1994) Script: Kurt Busiek / Art: Alex Ross

The greatest of all the *Avengers* story lines remains the Kree-Skrull War, in which Marvel super heroes first intervened in a war between galactic empires, as recalled here in *Marvels* over two decades after the story was first told.

non-Marvel super heroes of the period.

Thomas's most memorable saga during his six-year run on the *Avengers* was the Kree-Skrull War, between the two militaristic alien races created by Lee and Kirby in *The Fantastic Four*, with Earth and the Avengers caught in the middle. Running nine issues, much of it spectacularly illustrated by Neal Adams, the Kree-Skrull War had no precedent in comics; even Lee and Kirby had never devised a comic book epic of such scope. With this story *The Avengers* unquestionably established its reputation as one of Marvel's leading books.

THE VISION

Yet another of Roy Thomas's contributions to the *Avengers* was its most unusual member, the Vision. The original Vision was Aarkus, a green-skinned crimefighter from another dimension, who appeared in a short-lived series drawn by Jack Kirby in the early 1940s. Thomas wanted to create a new version for *The Avengers*. Since editor Stan Lee vetoed the idea of another extradimensional visitor, Thomas instead made the new Vision an android or, to use his term, "synthezoid," an artificial being composed of mock-organic parts and hence not a robot in the usual sense of the word.

Although he was a scientific invention, there seemed something supernatural about the Vision. He had the ability to control his own physical density mentally, enabling him to fly and to pass through solid objects. By increasing his density, he could become as hard as diamond and gain superhuman strength. Powered by solar energy, he could radiate intense heat from his eyes. He spoke in sepulchral tones (indicated by his characteristic word-balloon shapes) and his emotionlessness made him seem even more chilling. His most disturbing power was his ability to sink his immaterial hand into an opponent and then slightly rematerialize it, inducing intense shock in his victim; if the Vision made the least error, he could easily kill a person this way.

What made the Vision most intriguing were the hints of human emotions beneath his cold, robotic manner: even by the end of his second appearance in *The Avengers*, Thomas and his first artist, John Buscema, showed him weeping in gratitude at being accepted into the team. Writer Steve Englehart continued developing the Vision's emotions, allowing the Vision and the Scarlet Witch to fall in love with each other, to the outrage of bigots who refused to accept an android as the equal of a

Buscema's virile Vision
weeps in gratitude upon
being accepted into the
Avengers. Ever since then,
the android has rarely been
missing from active duty
with either the Avengers or
their West Coast team.

Avengers West Coast #50
(1988) Script and Pencils: John
Byrne / Inks: Mike Machlan

Following Thomas's suggestion,
later *Avengers* writer Steve
Englehart revealed that the
Vision was actually the reconstructed original Human Torch.
John Byrne disagreed with this
idea and, years later in the spin-
off series *Avengers West Coast*
disproved it by resurrecting
the Torch and having him shake
hands with the Vision, as the
Scarlet Witch, Wasp, and Wonder
Man watch in amazement.

human being—including Wanda's own brother
Quicksilver. The Vision and the Witch even got married, moved to a suburban home in Leonia, New
Jersey, and through magic were able to conceive
children, who were born at the conclusion of a
Vision and Scarlet Witch limited series by Englehart
and artist Richard Howell.

Writer/artist John Byrne felt a mistake had been
made and demonstrated the children to be merely
magical illusions, which then vanished for good. He
also broke up the Vision's marriage through a story
line in which the hero was abducted and disassem-
bled. Reconstructed and reprogrammed, the Vision
seemed even more emotionless than he had before.
Once again there is something inhuman and otherworldly about the Vision's manner, and he has
regained his previous mystique. At this writing his
emotions have once again begun to emerge and it
remains to be seen how they will develop this time
around.

MASTERS OF TIME

Roy Thomas was followed by Steve Englehart, who
did much of the best work of his writing career for
The Avengers from 1972 to 1976. His greatest
achievement was the story line that cast Mantis, a
Vietnamese bar girl and martial artist, as the
Celestial Madonna, the woman destined to bear a
child who would become the most powerful being
in the universe. Perhaps the true protagonist of this
story, however, was Kang the Conqueror, the most

compelling villain of *The Avengers* canon. Kang had first appeared in Lee and Kirby's *Fantastic Four* as a time traveler from the distant future who had gone back in time to ancient Egypt, where he used his advanced science to become the tyrannical Pharaoh Rama-Tut. Later, he traveled ahead to the fiftieth century A.D., where he became the warlord Kang and carved out an empire through time and space. He was eager to test his strength against the legendary heroes of the twentieth century, and he first clashed with the Avengers in their eighth issue.

Englehart had Kang seek to become the Celestial Madonna's mate and thus the father of her child of destiny. Englehart's stroke of brilliance was to bring Kang face-to-face with an incarnation of his own future self, a man who had tired of conquest and returned to ancient Egypt, this time to rule benevolently as Rama-Tut. This Rama-Tut now came to the present to thwart Kang's designs on the Madonna, but both were captured by another Lee-Kirby timelord, Immortus, who proved to be Kang's ultimate future self. By the end of his life Kang, as Immortus, had become a Prospero-like figure, living in hermetic isolation in his timeless limbo, manipulating both heroes and villains to achieve his mysterious ends. This story, with its shifting time frame and multiple identities, became one of the finer examples of Englehart's perennial theme of "the rising and advancing of a spirit" (as he put it in another of his series, *Master of Kung Fu*), a metaphor for the way a person takes on and casts off various aspects of personality as he or she moves through life.

WONDER MAN

Following Englehart's term on the series an old ally of the Avengers, long thought dead, Wonder Man, joined the team. Lee and Kirby intended Wonder Man to appear only in a one-shot story in *Avengers* #9; little did they know that later writers would revive virtually every character who appeared during Marvel's classic "Silver Age" years.

Simon Williams was a withdrawn scholar and inventor who found himself in over his head when he inherited his father's munitions business. Far outclassed by Tony Stark's innovations, Williams's business began failing, and the depressed industrialist embezzled company funds to try to save himself from bankruptcy. The theft was discovered and Williams was disgraced; deep in denial, he blamed not himself but Stark. Given his state of mind, Williams was easily manipulated by Captain America's enemy Baron Heinrich Zemo, who subjected him to an ionic radiation treatment that made him almost as strong as Thor and virtually indestructible but caused such severe side effects that he was dependent for his survival upon regular doses of a serum that only his benefactor possessed.

Williams, calling himself Wonder Man, agreed to infiltrate the Avengers, who were financed by the hated Stark, and betray them to Zemo. In order to win their trust, he came to their aid in battle and quickly gained membership in their ranks. He told them he was dying, and the Avengers generously strove to find him a cure. He soon led them into an

Avengers #8 (1964) Script: Stan Lee / Pencils: Jack Kirby / Inks: Dick Ayers

Kang the Conqueror, warlord from the far future, radiates confidence and ease as he first encounters his perennial foes, the Avengers.

Avengers #9 (1964) Script: Stan Lee / Pencils: Don Heck / Inks: Dick Ayers
At the beginning of his costumed career, Wonder Man "died" in saving the Avengers from Baron Zemo. Who suspected that he would be resurrected years later, having mutated into a being no longer physically human?

Avengers West Coast #54 (1990) Script and pencils: John Byrne / Inks: Paul Ryan
Transformed into a manifestation of pure "ionic energy" (and a part-time movie star), Wonder Man assisted his fellow West Coast Avengers in fighting the Mole Man's underground monsters in the streets of Los Angeles.

ambush, and Zemo captured the whole team. Wonder Man now had second thoughts: touched by their efforts to help him and ashamed of the depths to which he had descended, Wonder Man rejoined the Avengers. In the ensuing melee the Baron escaped and Wonder Man soon died.

And that seemed to be the end of Wonder Man. Indeed, Roy Thomas and John Buscema created a villain, the Grim Reaper, who was Simon's brother Eric and who spent years seeking revenge on the Avengers for Simon's death. Finally, however, the Reaper employed another villain, the Black Talon, to use voodoo to resurrect Wonder Man, whose near-indestructible body had remained intact. It turned out that Wonder Man had not truly been dead, but lay in a coma for years while his body continued its metamorphosis. Needless to say, he was welcomed back into the Avengers.

Williams proved to be a reluctant super hero. It did not matter that he seemed nearly invulnerable: nearly was not enough. Having "died" once, he was terrified of actually dying and felt like a coward and impostor fighting alongside his fearless colleagues. In any event, the life of a crusader was not enough for him. Curiously, considering his withdrawn childhood and scientific background, he chose to become an actor and kept taking leaves from the Avengers to pursue his career. After he moved west and became a successful stuntman, he felt confident enough to join the new West Coast Avengers. Before long Williams became an action-film star, a kind of B-movie Schwarzenegger.

Wonder Man finally received his own series, written by Gerard Jones as a satirical portrait of the film industry and Southern California life-styles, in which Simon Williams had to cope simultaneously with the insanities of Hollywood and super-villains. The parodic tone proved unsuccessful and the series turned to darker melodrama. In the course of these stories Williams learned that his fear of death was apparently moot: he was no longer a corporeal

THE BLACK KNIGHT

In the 1950s, perhaps influenced by Hal Foster's classic comic strip *Prince Valiant*, Stan Lee and artist Joe Maneely created *The Black Knight*, the saga of Sir Percy of Scandia (left), who posed as a foppish musician at the court of King Arthur. Secretly, though, he was the Black Knight, Camelot's greatest champion, who battled the conspiracies of Arthur's enemies Mordred and Morgan Le Fey and wooed the fair Lady Rosamund. His trademark weapon was the enchanted Ebony Blade, created by the wizard Merlin from a meteor.

The Black Knight lasted but six issues, but in the 1960s Roy Thomas introduced Sir Percy's American descendant, Dane Whitman, who assumed the role himself, wielding a lance that fired laser beams and riding a horse given wings through genetic engineering. Soon Whitman encountered Sir Percy's ghost and came into possession of the Ebony Blade itself. The present-day Black Knight has frequently served in the Avengers ever since then, although he has repeatedly been forced to contend with a blood curse placed on his enchanted sword. Below, Doctor Strange watches as he tries to purify the Ebony Blade.

Official Handbook of the Marvel Universe #13 (1984) Pencils: John Bolton / Inks: Josef Rubinstein

Doctor Strange (Second Series) #68 (1984) Script: Roger Stern / Pencils: Paul Smith / Inks: Terry Austin

Avengers #277 (1987) Pencils: John Buscema / Inks: Tom Palmer

One of the most chilling episodes of *Avengers* history was the work of writer Roger Stern, who followed Steve Englehart: the takeover of Avengers Mansion by Baron Zemo's vengeful son Helmut and a small army of the Avengers' old adversaries (right), beating even Hercules nearly to death before the Wasp led a contingent of Avengers in retaking their headquarters. In the aftermath Captain America learned that most of the few remaining links to his past had been destroyed (below). The follow-up was nearly as breathtaking, as the infuriated Zeus, wrongly believing the Avengers themselves responsible for Hercules' condition, imprisoned them in Pluto's underworld, from which they broke free to invade Olympus itself.

Avengers #277 (1987) Script: Roger Stern / Pencils: John Buscema / Inks: Tom Palmer

being but had become pure ionic energy in the form of a man. Soon after the *Wonder Man* series came to an end in 1993, Williams was apparently blown up in combat in *Force Works* #1, but considering that energy can be neither created nor destroyed, it seems safe to say that sooner or later he will be back.

THE EXPANSION TEAM

Numerous other super heroes, both stars of their own comics and featured players, have put in stints in the Avengers over the years. Among the female members have been the first Ms. Marvel and the She-Hulk; Tigra, a woman who had magically taken on catlike form and agility; the Black Widow, a former Soviet intelligence agent who became one of the team's leaders; the second Spider-Woman; Crystal of the Inhumans and Sersi of the Eternals; and even the Hellcat, who was really Patsy Walker, the longtime star of her own comic for teenage girls turned super hero. Other members included the first sorcerer hero of modern Marvel comics, Doctor Druid; the Beast of the X-Men, one of the Avengers' earliest foes; the Sub-Mariner; the resurrected original Human Torch; and even Mister Fantastic, the Invisible Woman, and the Thing at times when they had temporarily left the Fantastic Four. Even the mild-mannered but courageous Edwin Jarvis, the butler and majordomo of the Avengers' Manhattan mansion headquarters, has become a major character in the series over the years.

In the 1980s and 1990s *The Avengers* has continued to entertain its readers with the grandly scaled adventures they have come to associate with the title. Writer Jim Shooter and artist George Perez presented the mysterious Michael, an all-powerful being seeking to control the universe from his suburban backyard, who single-handedly held off the largest assemblage of Avengers to date. Recent years saw an attempt to surpass the scale of the Kree-Skrull War with a new intergalactic conflict, the Kree-Shi'ar War, a massive story line that spread from the Avengers into the individual members' own titles.

Reflecting the growing importance of the West Coast as decades passed—as well as its growing population of comics professionals—the Avengers founded a new division in the Los Angeles area, the West Coast Avengers, created by Roger Stern and Bob Hall, and launched into their own regular series by Steve Englehart and Al Milgrom. Although *Avengers West Coast* (as it was renamed) was finally canceled after a decade, several of its members immediately joined a new West Coast super hero team founded by Iron Man, Force Works, created by British writers Dan Abnett and Andy Lanning.

One recent *Avengers* story line, "Acts of Vengeance," brought the series full circle: Loki, rankled by the knowledge that his schemes had inadvertently created Earth's largest and mightiest band of super heroes, did his best to destroy them. But by this point *The Avengers* is unstoppable, a mainstay of the Marvel Universe that is sure to last as long as there is a Marvel Comics.

AVENGERS ASSEMBLE!

E ver since the Hulk quit in only the second issue, the Avengers' lineup has been continually in flux. Members come and go over the years, sometimes changing costumes or adopting new identities (like Henry Pym) as time passes. Most members who leave eventually return to the ranks. The one real constant presence in Avengers history, however, is that of their loyal and indispensable English butler, Edwin Jarvis

Avengers #71 (1969) Script: Roy Thomas / Pencils: Sal Buscema / Inks: Sam Grainger
The Avengers circa 1970 welcome the modern Black Knight into their ranks. Clockwise from the upper left, they include the Vision, Henry Pym as Yellowjacket, Hawkeye in his temporary role as the second Goliath, the Wasp, Captain America, Iron Man, the Black Knight, the Black Panther, and Thor.

MOMENTS LATER, IN A RATHER CLOSE ASSEMBLY ROOM...

TO BEGIN WITH, I THINK THE MAIN REASON WHY THE AVENGERS HAVE BEEN CONSIDERED A SECURITY RISK SHOULD BE PATENTLY OB-VIOUS--

--THERE ARE JUST TOO BLASTED *MANY* OF YOU! THE NATIONAL SECURITY COUNCIL CAN'T EVEN KEEP TRACK OF WHO'S COMING AND GOING!

Avengers #181 (1979) Script: David Michelinie / Pencils: John Byrne / Inks: Gene Day

The Avengers swelled to their greatest numbers during the war against Korvac, after which the federal government ordered a sharp downsizing of their membership. Also present in this panel are the Avengers' allies from an alternate future time line, the Guardians of the Galaxy. Included are the Beast, in between his periods serving in the X-Men; the Black Knight; the Black Widow; Captain America; Cap's frequent African-American partner, the Falcon, super hero of New York City's Harlem; Hawkeye; Hercules; Iron Man; Moondragon, an Earthwoman with vast psionic powers who was raised by the Eternals of the moon Titan; Ms. Marvel, alias Carol Danvers, a Captain Marvel supporting cast member turned super heroine in Marvel's attempt to create a feminist costumed heroine; Quicksilver; the Scarlet Witch; Thor; the Vision; the Wasp; Wonder Man; and Henry Pym in his guise of Yellowjacket. Also present is the Avengers' robotic ally Jocasta and the Guardians members Vance Astro, Charlie 27, Martinex, Nikki, Starhawk, and Yondu.

Avengers #397 (1996) Story: Terry Kavanaugh and Howard Mackie / Pencils: Mike Deodato, Jr. / Inks: Tom Palmer

The Avengers as of early 1996, with many in brand new costumes. From left to right: the new Swordsman (an ally), the Scarlet Witch, the Vision, Masque (another ally), Thor, butler Edwin Jarvis, Giant-Man, the Wasp, and Hawkeye.

--JUST SORRY WE MISSED OUT ON THE *REAL* ACTION.

'COURSE, IF OUR G.I. DANCE PARTNERS *WERE* REALLY UNDER ORDERS TO KEEP US OUT OF THE BATTLE UP NORTH, LIKE THEY CLAIMED...

THEN IT'S MORE THAN LIKELY THAT SOMEONE IN THE CHAIN-OF-COMMAND HAS BEEN *COMPROMISED* BY THIS SO-CALLED OMNIBUS.

AND THE *ONLY* ONES WHO CAN CONFIRM THAT--WHO CAN *POINT* US TO THIS OMNIBUS CHARACTER--

Gene
Colan

FIVE

STRANGE TALES: HEROES OF THE SUPERNATURAL

Marvel writers concoct science-fictional explanations for their characters' amazing abilities, but in truth, these powers are really the equivalents of the magical abilities of the demigods and wizards of myths, legends, and fables of past centuries. In twentieth-century America the wonders of science seem magical, and so pseudoscience takes the place of sorcery in much of our adventure fantasy.

But not in all of it. There is still an audience for the traditional archetypes of supernatural horror and fantasy: the magician, the vampire, the manlike beast lurking in the wild. Over the years, Marvel's writers and artists have found ways to transport these archetypes into the contemporary Marvel Universe.

Doctor Strange (Second Series) #56 (1982) Script: Roger Stern / Pencils: Paul Smith / Inks: Terry Austin

The doctor is in: Stephen Strange, Sorcerer Supreme of the Marvel Universe.

DOCTOR STRANGE

The foremost of Marvel's supernatural protagonists is Doctor Strange, Earth's sorcerer supreme, created in 1963 by Stan Lee and Steve Ditko to star in a back-up feature for *Strange Tales*, which then featured solo adventures of the Fantastic Four's Human Torch. Strange stands in the pantheon of Marvel's most important heroes, even though his popularity among the mass audience never matched that of Lee and Ditko's other major creation, Spider-Man. Over the decades he has starred in two series of his own that were canceled; the third, *Doctor Strange, Sorcerer Supreme*, continues to this day.

Perhaps Strange has remained a cult favorite because his abilities do not have the same visceral power as those of most Marvel super heroes: he uses neither his fists nor a weapon but instead casts rhymed spells. Nor does he suffer the angst of being young or an outsider like so many other Marvel heroes; indeed, he is very clearly a middle-aged man. Strange's appeal is quite different: he is a cerebral figure and, like Lee's Silver Surfer, a spiritual one. Strange's best stories exemplify and celebrate the power of the imagination.

Stephen Strange was once the embodiment of pride and greed, an arrogant surgeon caring more for the size of his fees than the health of his patients. Brought low by an auto accident that permanently injured his hands, leaving them too unsteady to perform delicate operations, and consumed by despair, he lost everything, eventually declining into drunken dereliction. Hearing sailors speak of miraculous cures performed by a man known as the "Ancient One," Strange journeyed to India (changed in later retellings to Tibet) to seek this mystic who might heal his hands.

Strange spent years with the Ancient One training to become a "master of the mystic arts" and then returned to America, settling into an unusual Victorian-style house on Bleecker Street in bohemian Greenwich Village in New York City. He was regarded by the world at large, which did not believe in magic, as an eccentric and a charlatan. But in fact he acted as mankind's point man, the sole defender of humanity from perils it did not imagine existed.

What differentiated Strange from other fictional magician heroes was that he did not primarily operate in the real world. His foes were not common criminals and, with the notable exception of his archrival Baron Mordo, often not even other human sorcerers. They were, rather, beings who inhabited

I SEE YOU IN THE PAST...IN AMERICA...YOU ARE WEARING THE FROCK OF A *DOCTOR!* AHH, YOU WERE A FAMOUS SURGEON NAMED *STEPHEN STRANGE!*

"YOU WERE PROUD, HAUGHTY, SUCCESSFUL! BUT YOU CARED LITTLE FOR YOUR FELLOW MEN..."

THE OPERATION WAS A SUCCESS, DOCTOR! YOUR PATIENT WANTS TO THANK YOU!

I CAN'T BE BOTHERED! JUST BE SURE HE PAYS HIS BILL!

"MONEY...THAT WAS ALL THAT INTERESTED YOU...ALL YOU CARED ABOUT..."

SORRY, IF YOU WON'T PAY MY PRICE, I CAN'T HELP YOU! FIND ANOTHER DOCTOR!

THE ANCIENT ONE!! MANY TIMES IN THE PAST I, TOO, HAVE HEARD THIS NAME MENTIONED IN LOW WHISPERS! CAN IT BE THAT THERE IS SOME *TRUTH* TO THE LEGENDS? HISTORY TELLS US THERE *HAVE* BEEN MEN WITH CERTAIN POWERS...WHAT IF *HE* IS SUCH A MAN?

THE REST IS EASY TO DEDUCE! YOU SOUGHT ME FOR MY HEALING POWER! BUT I CANNOT HELP YOU...FOR YOUR MOTIVES ARE STILL SELFISH!

AND YET, I SEEM TO SEE A SPARK WITHIN YOU...A SPARK OF DECENCY...OF GOODNESS..WHICH I MIGHT BE ABLE TO FAN INTO A FLAME!

IF YOU WILL STAY HERE...STUDY WITH ME...PERHAPS YOU WILL FIND WITHIN YOURSELF THE CURE YOU SEEK!

I SHOULD HAVE *KNOWN!* IT WAS JUST A WASTE OF TIME! YOU'RE NOTHING BUT AN OLD FRAUD!

I AM ONLY SUBJECT TO MORDO'S *SPELL* IF I TRY TO *WARN* THE ANCIENT ONE! YET, I AM ABLE TO SPEAK OF *OTHER* MATTERS...SO THERE IS STILL ONE HOPE!

IF *I, TOO,* CAN LEARN THE SECRETS OF BLACK MAGIC, THEN *I* CAN BATTLE MORDO WITH HIS OWN WEAPONS!

ANCIENT ONE, I CRAVE A BOON! I WISH TO ACCEPT THE TERMS YOU OFFERED ME SOME DAYS AGO! I WISH TO STUDY AT YOUR FEET...TO BE TAUGHT YOUR KNOWLEDGE...TO PROVE MYSELF WORTHY OF THE MYSTIC ARTS!

AHH! AT LAST I HAVE REACHED THE *REAL* DR. STRANGE!

I KNEW THAT THERE WAS *GOOD* WITHIN YOU...IF I COULD BUT BRING IT TO THE SURFACE! I *ACCEPT* YOU, MY SON! YOU SHALL BE MY DISCIPLE!

Stephen Strange was an arrogant man of the world (top) when an automobile accident abruptly altered his destiny. Without his manual skills, he quickly descended from the heights of fame and fortune to the lower depths, becoming an alcoholic tramp (center). Like many skeptics throughout the ages, he accepted the existence of the spiritual realm only after witnessing a "miracle": while visiting the stronghold of the Ancient One, he secretly watched as the Ancient One's disciple, Baron Mordo, prayed to his dark god Dormammu for the power to slay the old mystic. Discovering Strange, Mordo used a spell to clamp his mouth and bind his wrists to prevent him from warning the Ancient One. Amazed by his discovery that sorcery is real and shocked by the depth of Mordo's evil, Strange felt a new resolve to do whatever he could to stop the Baron. He accepted the Ancient One's offer to become his pupil (bottom) and "prove myself worthy of the mystic arts!"

"You have been tested, and you have passed your baptism of fire!" said the old man. "But the path ahead of you will be difficult and fraught with danger! So you still wish to continue?"

"I do," Strange replied.

Strange Tales Vol. 1 #115 (1963)
Script: Stan Lee / Art: Steve Ditko

SURREALISM

Steve Ditko's artwork was crucial to the evolution of the series. He created uniquely surreal landscapes for these other worlds, with odd, geometric forms floating in the void, and equally unusual denizens for these realms, entirely different in appearance from the standard alien beings populating other planets in the comics of the period.

Strange Tales Vol. 1 #138 (1965) Plot and art: Steve Ditko / Script: Stan Lee

Strange Tales Vol. 1 #122 (1964)
Script: Stan Lee / Art: Steve Ditko

Before there was *A Nightmare on Elm Street*'s Freddy Krueger, there was Doctor Strange's enemy Nightmare, a being who can manipulate the world within our dreams at will.

Strange Tales Vol. 1 #140 (1966)
Script: Stan Lee / Art: Steve Ditko

Strange's supreme archenemy, the dread Dormammu, challenges him to a duel before an audience composed of the rulers of other mystical dimensions.

other "dimensions" where magic, not science, ruled, and who, if not stopped, would turn humanity's realm of order and rationality to utter chaos.

One of the leading figures in Lee and Ditko's pantheon of mystical entities was introduced in Strange's very first story: Nightmare, a thin, ghastly, pale-skinned figure mounted upon a jet black steed. Like a character in a medieval allegory, he embodied his name: he was indeed the spirit of nightmare, menacing individuals through their subconscious minds. Strange entered his dimension by projecting himself into the dream of one of Nightmare's victims or by dreaming himself. In combating Nightmare, Strange was literally fighting his own vulnerability to his buried fears: should he fall asleep unprotected by spells he too became Nightmare's victim. (Nightmare was the first of what writer Mark Gruenwald would later dub the "conceptual beings" of the Marvel Universe: individual characters who embodied aspects of the universe or of existence, like Death itself, first encountered by Strange and

I HAVE HEARD FATHER SPEAK OF AN ANCIENT ONE WHO BATTLED DORMAMMU LONG AGO.! IT CANNOT BE THE SAME ONE!

FOR *HE* IS YOUNG-- AND FAIR TO BEHOLD!

Strange Tales Vol. 1 #126 (1964)
Script: Stan Lee / Art: Steve Ditko

Clea, deposed princess of Dormammu's Dark Dimension, first catches sight of her future lover, Stephen Strange.

Strange Tales Vol. 1 #127 (1964)
Script: Stan Lee / Art: Steve Ditko

Even Dormammu feels threatened by the Mindless Ones, living embodiments of unrestrained brute violence.

YES.! AND THERE THEY *ARE*...PRIMITIVE, SAVAGE, TOTALLY DEVOID OF LOVE, OR KINDNESS, OR ANY TYPE OF INTELLIGENCE! THEY LIVE ONLY TO *FIGHT*... AND TO DESTROY!

▲ *Doctor Strange, Sorcerer Supreme* #21 (1990) Script: Roy and Dann Thomas / Pencils: Jackson Guice / Inks: Mark McKenna

A family portrait: Years passed before it was revealed that Clea, who grew more glamorous over the years, was actually Dormammu's niece.

Slowly, the figure speaks ... in a voice which is not a voice ... mouthing words which are more than words ... expressing thoughts no mortal has ever gleaned before!

I AM ETERNITY! HEED MY MESSAGE AND REMAIN SILENT, FOR NONE MAY SPEAK WHEN I AM PRESENT!

YOU ARE THE SECOND MORTAL TO STAND BEFORE ME! THE FIRST WAS HE WHOM YOU SERVE ... KNOWN TO YOU AS THE ANCIENT ONE! IT WAS TO HIM I GAVE THE MAGIC AMULET, AND THE POWER IT CONTAINS!

Strange Tales Vol. 1 #138 (1965) Plot and art: Steve Ditko / Script: Stan Lee

An early high point of the series: Strange passes through the mind of the frail Ancient One to enter a reality in which he encounters Eternity himself. This was perhaps Ditko's most unforgettable and awe-inspiring visual creation: Eternity, the living embodiment of all of time and space, pictured as a gigantic humanoid outline through which stars and galaxies shone, with an impassive half-mask serving as his face.

▶ **Strange Tales** Vol. 1 #157 (1967) Script: Stan Lee / Pencils: Marie Severin / Inks: Herb Trimpe

The Living Tribunal is a single being with three, inhuman, passionless faces, one wholly covered by his veil, each representing a different aspect of justice. Perhaps the most powerful entity in Strange's series, the Tribunal guards all of reality against imbalances of mystical power, even if it should mean destroying Earth.

Marvel Premiere Vol. 1 #14 (1974) Script: Steve Englehart / Pencils: Frank Brunner / Inks: Dick Giordano

In response to humanity's evil, the evolving deity Sise-Neg devastated Sodom and Gomorrah on one of his stops through time. But on Strange's urging, he also stopped to create a paradisiacal garden for a protohuman and his mate.

later the muse of the evil alien Thanos in Jim Starlin's stories for Captain Marvel and Warlock.)

The archvillain of Strange's cosmos was the dreaded Dormammu, dictator of the Dark Dimension, a virtual god of evil, unforgettably visualized by Ditko as a humanoid being with a raging fire in place of his head. It was in the Dark Dimension that Strange met and paid court to its rightful princess, Clea. Over the years Clea, who began as the helpless victim of Dormammu and his equally sinister sister Umar (eventually revealed to be Clea's mother), has come into her own as a character. She came to Earth to become Strange's lover and apprentice and eventually grew into an indefatigable revolutionary leader seeking to overthrow her tyrannical mother and uncle.

Though Strange could not overpower the far mightier Dormammu, in their first encounter they reached a stalemate when they had to cooperate against another of Ditko's startling visual creations: the raging, cyclopean Mindless Ones that forever storm the mystic barrier sealing them off from the Dark Dimension. Ultimately Strange and Dormammu met in combat before an audience composed of the rulers of other mystical dimensions: both duelists renounced the use of their mystical abilities for the battle, instead wielding only "the pincers of power." After a harrowingly depicted

Doctor Strange (First Series) #177 (1969) Script: Roy Thomas / Pencils: Gene Colan / Inks: Tom Palmer

In the late 1960s writer Roy Thomas and artist Gene Colan sought to make Doctor Strange more appealing to the rest of Marvel's readership by making him look more like a super hero. Doctor Strange's first solo comic book was canceled anyway, and he discarded his new look when his series was revived.

fight, the mortal overcame the immortal, forcing Dormammu to confess defeat before his peers.

As with *Spider-Man*, Ditko's run as co-plotter and artist on *Doctor Strange* was all too short, but in the latter's case he built to a powerful denouement, in which, before the horrified eyes of the mortal Strange, Dormammu attacked Eternity, causing a cataclysmic burst of power in which both entities vanished.

Stan Lee continued to write classic *Strange* stories after Ditko's departure, collaborating with artists Bill Everett, the first to draw the seductive Umar, and Marie Severin, who visualized one of the most remarkable entities in the Strange canon: the giant, hooded Living Tribunal, who first confronted Strange amid the eerie surroundings of Stonehenge.

It was writer Steve Englehart and his collaborators, artists Frank Brunner and Gene Colan, who, in the 1970s, did the most innovative and memorable work with the character after Lee's own stories. Englehart and Brunner chronicled Strange and Mordo's pursuit of a sorcerer from the future, Sise-Neg, as the latter traveled backward through time, absorbing extraordinary amounts of mystical power. The two enemies found themselves in roles comparable to those of the good and bad angels of medieval drama, Strange trying to persuade Sise-Neg to use his vast power for good while Mordo played the tempter. Ultimately the trio reached the moment of the creation of the universe, when, in effect, Sise-Neg merged with God, achieved transcendence, saw the value in all of the creation he had witnessed, and willed time to begin anew.

Many writers and artists have contributed to the legend of Doctor Strange since Lee's original stories: Roy Thomas, Roger Stern, Peter B. Gillis, Barry Windsor-Smith, Michael Golden, Marshall Rogers, Paul Smith, and Jackson Guice, to mention only a noteworthy handful. In late 1995, J. M. DeMatteis inaugurated his run as Strange's new chronicler with a story line in which the Doctor solves a mystery by unearthing buried memories from his own childhood. As ever, Strange's most profound quests are those he makes, literally or symbolically, into his own soul.

SILVER DAGGER

Even more remarkable than the story of Sise-Neg was the saga of the religious fanatic turned witch-hunter, Silver Dagger (below), who slew Strange and took Clea captive. Strange's spirit descended into his mystical "crystal ball," the Orb of Agamotto (left), where he found a surreal world, mixing distorted mirror images of his friends with horrors from his subconscious, and, memorably, the mystical deity Agamotto himself in the guise of a taunting, talking caterpillar out of Lewis Carroll (below left). To leave this world of unreality Strange had to face and submit to Death, manifested as an actual mystical being. By this means Strange transcended his own mortality, enabling him to return to the real world and to Clea, joining with her (briefly even inhabiting her body in a fine image of the bond between them) to overcome Silver Dagger; Strange returned to life, more powerful and wiser than before, while his foe found himself exiled to a hell of unreality within the Orb.

Doctor Strange (Second Series) #1 (1974) Script: Steve Englehart / Pencils: Frank Brunner / Inks: Dick Giordano

Doctor Strange (Second Series) #1 (1974) Script: Steve Englehart / Pencils: Frank Brunner / Inks: Dick Giordano

Doctor Strange (Second Series) #5 (1974) Art: Frank Brunner

Doctor Strange, Sorcerer Supreme #6 (1989) Script: Roy and Dann Thomas / Pencils: Jackson Guice / Inks: Jose Marzan, Jr.

Doctor Strange, Sorcerer Supreme #6 (1989) Art: Jackson Guice

Artist Jackson Guice conjured a mixture of the beautiful and the macabre during his stint drawing Doctor Strange, as in these two portraits of Mephista (above and above right), the daughter of the Silver Surfer's demonic archfoe Mephisto.

Doctor Strange, Sorcerer Supreme #84 (1995) Art: Mark Buckingham

In the mid-1990s Doctor Strange has been given an elaborately designed costume in a new attempt to keep up with the times.

Tomb of Dracula #29 (1975)
Script: Marv Wolfman / Pencils:
Gene Colan / Inks: Tom Palmer

Colan and Palmer superbly conjured
the atmosphere of classic horror
within contemporary times, locating
their stories in the haunted English
countryside and the darkened streets
and centuries-old edifices of London
and Boston, milieus where men and
women in modern dress could somehow credibly cross paths with a
demonic figure in the garb of a
previous century.

DRACULA COMES TO MARVEL

As a response to congressional investigations of comics in the 1950s, the comics industry had formulated the Comics Code Authority, which forbade "horror" comics such as EC's once notorious and now classic *Tales from the Crypt*. In the early 1970s the Code's prohibition on this genre was relaxed, and Marvel launched numerous comics based on traditional horror themes, such as *Werewolf by Night*, which was actually inspired by the 1957 B-movie *I Was a Teenage Werewolf*, and *The Monster of Frankenstein*, which began with a retelling of Mary Shelley's novel brilliantly illustrated by Mike Ploog.

The flagship of these titles, however, was *Tomb of Dracula*, which outlasted all the rest, running for seventy issues. Various writers came and went during the comic's initial year until writer Marv Wolfman joined the creative team of penciller Gene Colan and inker Tom Palmer; together they made *Tomb of Dracula* one of the great comics of the 1970s.

The overarching plot line of *Tomb of Dracula* had a small band of vampire hunters seeking at every turn to thwart Dracula's depredations upon humanity. But action, suspenseful as it often was in *Tomb*, was not truly the heart of the series; rather, the stories in *Tomb of Dracula* were driven by its writer's and artists' skilled delineation of character.

Some of today's comics writers might be tempted to portray Dracula's nemeses as latter-day Punishers bearing stakes instead of assault weapons. Wolfman and Colan instead made their

Tomb of Dracula #13 (1973)
Script: Marv Wolfman / Pencils:
Gene Colan / Inks: Tom Palmer

Vampire hunters, clockwise from the
left: Quincy Harker, flighty receptionist
Aurora Rabinowitz, Rachel Van
Helsing, Dracula's descendant Frank
Drake, and Woody Allen wanna-be
Harold H. Harold.

Tomb of Dracula #41 (1976)
Script: Marv Wolfman / Pencils:
Gene Colan / Inks: Tom Palmer

As Harker, Drake, and Van
Helsing watched, Blade finally
impaled Dracula with one of his
trademark wooden knives; as
usual, however, Dracula did not
remain dead for very long.

band of vampire hunters a group of credible, vulnerable human beings capable of introspection, of
doubt, and of warmth toward others, who could lead
happy, fulfilling lives were it not for Dracula's intrusion. One of the band, Frank Drake, was Dracula's
sole twentieth-century descendant, who had made a
useless botch of his life until the hunt for Dracula
gave him purpose. He fell in love with the woman
who brought him into the group, Rachel Van Helsing,
the granddaughter of Dracula's nemesis from Bram
Stoker's original novel. As a child, Rachel had seen
her parents murdered by Dracula; scarred physically
and emotionally, she became a formidable hunter
driven by vengeance, but facing an even greater
struggle in trying to feel trust and love again.

Acting as a kindly father figure to both Frank
and Rachel was Quincy Harker, the son of another
of Stoker's characters. Now an elderly man bound
to a wheelchair, Quincy had lost virtually everything
to Dracula: his wife and daughter, even the mobility
in his legs. To Harker the war with Dracula sometimes seemed wearyingly endless and futile: Quincy
was old, approaching death, while his nemesis
seemed immortal. Nonetheless, Harker bravely continued the fight.

Among their allies was Blade, an African-
American wielding wooden blades, who sought the
vampire who killed his mother as she gave birth,
leaving him immune to vampire bites. He formed a
reluctant partnership with Hannibal King, a detective turned vampire who spoke in dialogue straight
out of a Raymond Chandler mystery; refusing to

drink human blood or give in to his curse in any way,
he fought Dracula's legion of vampires from within.

Wolfman and Colan's greatest success in characterization was Dracula himself. They skillfully
molded what could have been a cardboard villain
into a three-dimensional figure. Marvel's Dracula
was less an outright devil than a fallen angel, a
noble being turned to evil: genuinely grand, even
regal; evil yet respectful of his adversaries and given
to unexpected acts of kindness to strangers; vicious
and cunning yet capable of genuine love.

In the series' final and most stunning story line

Tomb of Dracula #55
(1977) Script: Marv
Wolfman/Pencils: Gene
Colan/Inks: Tom Palmer

In one of *Tomb's* most
intriguing story lines
Dracula fell in love with
Domini, a young woman
who through mystical
means bore him a son.
Ironically, the child was
possessed by an angel, who
transformed his host into
an adult, sworn to destroy
the vampire lord. What
was remarkable was
Dracula's reaction: not simply rage at his heavenly
foes, but anguish that even
his own child had been
taken from him. Despite
her desperately bad taste
in husbands, Domini was a
rather saintly woman, and
Colan's portrayal here
evokes the image of a
Renaissance Madonna.

Tomb of Dracula #32
(1975) Script: Marv
Wolfman/Pencils: Gene
Colan/Inks: Tom Palmer

The aged Quincy Harker,
son of the hero of Bram
Stoker's novel, waged a
lifelong war against
Dracula that ended when
Harker set off an explosion
that killed both himself
and—temporarily, of
course—the vampire lord.

Dracula quarrels with a demon purporting to be
Satan, who then removes his vampiric powers,
reducing him to the level of a homeless, pathetically
vulnerable human, still pursued by the enemies he
made in the past; the demon restores Dracula's
powers only after driving him to the breaking point,
which in Dracula's case means calling upon God for
mercy. Humiliated, Dracula returns to Transylvania,
defeating a pretender to his throne as vampire lord,
before engaging in a final confrontation with
Harker. The two old enemies perish together, and
Drake and Van Helsing are left tentatively to attempt
to find a life with each other now that their hunt has
finally ended. Various Marvel creators have resurrected Dracula for further tales, but none of them
have managed to equal or surpass the dramatic
power of *Tomb of Dracula* at its peak.

THE MAN-THING

The most extraordinary of Marvel's "horror" series of
the 1970s was the *Man-Thing*, which began in
Savage Tales and continued in *Adventure into Fear*
and his own titles. The character was the inspiration
of Roy Thomas, as his tribute to science-fiction
writer Theodore Sturgeon's "It" and the 1950s
comic book variant, the Heap; both creatures were
men who died in the waters of a swamp and
returned to a parody of existence as virtually mindless beings composed of vegetable matter, driven by
primal memories from their past. (Almost simultaneously with Man-Thing's debut, DC came up with its
own version of the Heap, Swamp Thing.)

The first *Man-Thing* story, written by Gerry
Conway, introduced Ted Sallis, a scientist who was
working in an isolated cabin in the Everglades to
re-create the "super-soldier serum" that had turned
Steve Rogers into Captain America. Like Bruce
Banner, Sallis cared nothing about the potentially
immoral uses to which the military might put his
serum; this was the sin that brought on his fate.
Betrayed by his girlfriend to spies, Sallis injected
himself with the serum to prevent them from getting it. In the ensuing struggle, Sallis was hurled
into the swamp, where the serum interacted with
the murky water and muck, causing him to mutate
into a mindless, ghastly monster.

The Man-Thing is man reduced to his essentials,
and then reduced even further than that. He is an
empath, one who can sense the emotions of others

Giant-Size Man-Thing #1 (1974)
Script: Steve Gerber/ Pencils: Mike Ploog/ Inks: Frank Chiaramonte
The elephantine Man-Thing wanders aimlessly through his swamp in the Everglades, his life devoid of conscious purpose.

and is pained by the presence of hatred, violent anger, and, worst of all, fear. The Man-Thing will act to protect the innocent, although clearly not knowing why. But on sensing strong negative emotions, the Man-Thing will attempt to put an end to them, even if it means killing the perpetrator; as the narrator solemnly states in each story, "Whoever knows fear burns at the Man-Thing's touch."

Since the main character is virtually incapable of thought, his stories actually center around the people he encounters; hence, this is one Marvel series that by its very nature focuses on characterization rather than action. It was writer Steve Gerber who seized the possibilities *Man-Thing* presented and made it a classic. Together with his various artists, notably Val Mayerik and Mike Ploog, Gerber turned Man-Thing's swamp into a phantasmagorical stage for his imagination.

He concocted strange sagas of love and hatred in relations between the sexes. A comatose woman somehow transforms trees and animals into monsters through the force of her subconscious hatred for her spouse. An arrogant modern woman, Dr. Maura Spinner, is abducted by Captain Fate, a pirate of centuries past whose ship now roves another dimension; she becomes a pirate queen but learns the value of love from her encounter with a grotesque satyr named Khordes. Among Gerber's most moving stories were "Song-Cry of the Living Dead Man," about Brian Lazarus, an advertising

Giant-Size Man-Thing #4 (1975) Script: Steve Gerber/ Pencils: Ed Hannigan/ Inks: Frank Springer
Deeply empathic, the Man-Thing experiences pain in the presence of dark emotions such as hatred or the fear that underlies it. To stop his suffering, the Man-Thing seizes its source, triggering a chemical reaction that causes the flesh of anyone who feels fear to burn as if exposed to a powerful acid.

Giant-Size Man-Thing #3 (1975) Script:
Steve Gerber/Art: Alfredo Alcala

THE NEXUS OF ALL REALITIES

One never knew what to expect from Steve Gerber's *Man-Thing* stories; they could take the reader virtually anywhere. Indeed, Man-Thing's swamp proved to be the "Nexus of all Realities," a gateway to other universes. One supporting character, the teenager Jennifer Kale, was a seemingly ordinary girl living on the edge of the swamp who discovered her talent for sorcery. As the series went on, she found herself journeying to other dimensions with a motley array of fellow heroes, including the Man-Thing, a warrior swordsman named Korrek (whose name is misspelled in this panel's caption), an eccentric sorcerer named Dakimh, and, briefly, a talking duck named Howard, to prevent the takeover of all reality by a demon named Thog who masqueraded as a yuppie businessman. Another recurring character was Richard Rory, a young disc jockey cursed by ill fortune and low self-esteem, a man of the 1960s trying to make his way in the less optimistic 1970s, who discovered inner resources he feared he lacked. Among his adversaries was the original Foolkiller, a religious fanatic armed with a gun that cast the "light of purification" to obliterate all those he judged to be life's "fools." In the swamp, Rory and the Man-Thing witnessed the trial of the spirit of a dead clown by representatives of heaven and hell, consisting of a review of episodes of his life to determine whether or not it rose sufficiently above mediocrity to be worthy of either eternal bliss or damnation.

copywriter driven to insanity by the demands of his life, whose hallucinations take on reality; "A Candle for Sainte-Cloude," in which Ted Sallis briefly encounters a young, free-spirited woman who could have turned him from the path that led him to becoming the Man-Thing; and "The Kid's Night Out," which begins with the students and faculty of a high school mourning the recent death of a youth who went there, and follows his lone friend as she exposes the hypocrisy of this little community, showing how the boy was mocked, persecuted, and finally literally beaten to death for the supposed sins of being overweight and unpopular.

Through it all the Man-Thing stood as judge, watching the passing events with passive curiosity, but intervening at the climactic moment of the drama, moved by his primal revulsion toward evil to mete out retribution to the perpetrators of injustice, who invariably proved to be cowards vulnerable to his burning touch.

Man-Thing was at least fifteen years ahead of its time, a forerunner of many contemporary comics that use the framework of supernatural fantasy to explore human psychology and emotion. Gerber's *Man-Thing* was also an intensely personal work; appropriately, in the original series' final issue Gerber portrayed himself coming face-to-face with Man-Thing. Neither the subsequent attempts to revive the Man-Thing nor even most of the supernatural fantasy stories of today's comics come anywhere close to creating such a powerful sense of the writer speaking so intimately, dramatically, and directly to his reader.

HOWARD THE DUCK

The failure of the 1986 *Howard the Duck* movie has obscured the real virtues of the comic book series on which it was based. Steve Gerber's *Howard the Duck* comic was a broad parody of fantasy, science fiction, and super hero stories, but it was also an insightful satire of America in the 1970s.

Born on an other-dimensional Earth populated by talking ducks, Howard was transported to Man-Thing's swamp by his creators Gerber and artist Val Mayerik in 1973 and soon appeared in stories of his own. Like many Marvel super heroes, he is an outsider; being a duck in a human society proved to be both funny and strangely poignant. Howard acts like a cynical, embittered adult, but he is also like a child, abruptly bereft of the security of home, finding himself in an adult world where everyone is taller than he, refuses to take him seriously, and insists he follow their rules. Howard reacts by taking apart the target of his outrage with some pointed observations and a disgusted quack (right). With a hero who was the archetypal outsider able to see through the absurdities of the society around him, *Howard the Duck* gave Gerber a platform for an ongoing critique of contemporary American fools and pretenders, from the power-mad capitalist wizard Pro-Rata to the cult leader Reverend Joon Moon Yuc.

Howard the Duck #1 (1976) Script: Steve Gerber / Pencils: Frank Brunner / Inks: Steve Leialoha

Howard the Duck #6 (1976) Script: Steve Gerber / Pencils: Gene Colan / Inks: Steve Leialoha

Howard the Duck even challenged the very foundation of the action-adventure genre. During his celebrated run for the U. S. presidency in 1976, Howard said he was in favor of violence in entertainment media, "as long as it's never presented as cathartic—as a release, as a solution." Howard himself avoided fights whenever possible.

The heart of the series lay in Gerber's depiction of the relationship between Howard and his human companion, Beverly Switzler (left). Their bantering rapport, quarrels, and reconciliations were funny but also had an unexpectedly realistic dimension. The bond between this unlikely pair seemed a metaphor for the power of love to bridge the gap between individuals, no matter how different they are from one another.

Official Handbook of the Marvel Universe #14 (1984) Pencils: Bob Hall / Inks: Josef Rubinstein

ZOMBIE

Steve Gerber found a character like the Man-Thing in Simon Garth, the ruthless New Orleans businessman who was murdered by an employee and raised from the dead as the eponymous lead of Marvel's 1970s black-and-white comics magazine *Tales of the Zombie*. Like the Man-Thing, Garth, having lost most of his memory and reasoning abilities, was now a bundle of primal needs; pathetic and horrifying as his fate was, he had also somehow been purified of the evil of his past existence. Unlike the Man-Thing, Garth was allowed to find salvation: in a story by Tony Isabella he was magically restored to life in order to spend a day making up for his past sins before he reverted to his undead state and was laid finally to rest. (Or, rather, about as final as most Marvel deaths get, since two decades later *Daredevil*'s creative team saw fit to resurrect him as a zombie yet again.)

GHOST RIDER

The most successful of the 1990s supernatural series actually began two decades earlier. Created by Gary Friedrich and Mike Ploog, the Ghost Rider of the 1970s took the name of Marvel's wraithlike hero—now known as the Phantom Rider— from its line of Westerns. In appearance he was surely inspired by an obscure Marvel/Timely hero of the 1940s, the Blazing Skull.

Stunt motorcyclist Johnny Blaze made a deal with the "devil" (later identified as the Silver Surfer's nemesis Mephisto) to save the life of his stepfather, "Crash" Simpson, who was dying of cancer. Mephisto cured the cancer but then saw to it that Simpson died in an accident. To fulfill his end of the deal, Blaze was repeatedly transformed into a spectral figure, with a fiery skull for a head, clad in Blaze's leather outfit, and riding a blazing motorcycle. The Ghost Rider Blaze altruistically fought both demons and criminals alike, but he underwent a personality change as the series progressed, becoming increasingly ruthless and inhuman in his transformed state. It finally was revealed that the Ghost Rider was actually Zarathos, a demon enslaved by Mephisto, using Blaze as a living human host. Blaze rid himself of Zarathos in the final issue of the series, which ran for many years, and returned to a normal life.

A new version of *Ghost Rider*, written by Howard Mackie, proved wildly successful on its debut in 1990. It began with teenager Dan Ketch

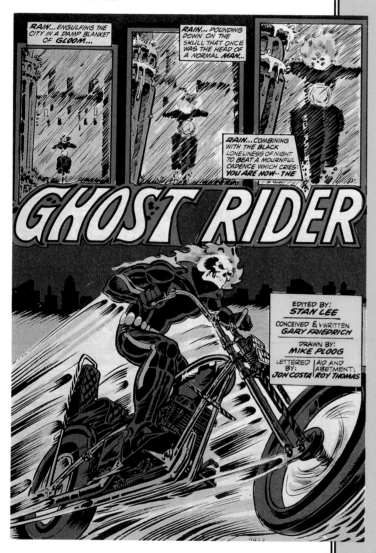

Marvel Spotlight #5 (1972) Script: Gary Friedrich / Art: Mike Ploog
Possessed by the demon Zarathos, stunt motorcyclist Johnny Blaze became the first Ghost Rider.

and his sister Barbara coming across a meeting of criminals one night in New York City's Cypress Hills Cemetery. Not wanting any witnesses, one of the criminals shot Barbara down. Desperately determined to stop them from harming Barbara any further, Ketch laid hold of a strange abandoned motorcycle and was transformed into a new Ghost Rider, an updated image of Blaze's Ghost Rider of the 1970s. This new Ghost Rider proclaimed himself a Spirit of Vengeance out to punish those who

HELLSTORM

Steve Gerber was the principal writer for another 1970s series, *The Son of Satan*, whose strange hero, the exorcist and demonologist Daimon Hellstrom, was the son of a mortal woman and a demon purporting to be Satan. (Usually, Marvel has steered clear of actually portraying God or Satan.) Discovering his demonic heritage, Hellstrom wielded a devil's trident from which he projected "hellfire," but turned it against the forces of the underworld. Hellstrom was revived in the 1990s in a series bearing his new name, *Hellstorm*. Now portrayed as a grim, foreboding figure, Hellstrom, in stories by British writer Warren Ellis, claimed that both heaven and hell were the enemies of humanity and set himself in opposition to both; by the series' end he had displaced his father as master of hell.

Hellstorm #15 (1994) Art: Derek Yaniger

Morbius #1 (1992) Pencils: Ron Wagner / Inks: Mike Witherby

MORBIUS

One of Marvel's supernatural antiheroes did not technically fit that description at all. The scientist Michael Morbius concocted what he hoped was a cure for his rare blood disease that instead transformed him into a "living vampire," a creature whose ghastly appearance, superhuman abilities, and need to ingest blood resembled the traits of "real" supernatural vampires. Morbius first appeared in *Amazing Spider-Man* #101 and was clearly an attempt by his creators, Roy Thomas and Gil Kane, to introduce an element resembling the supernatural into a series in which magic—apart from guest appearances by Doctor Strange—played no part. What most clearly distinguished Morbius from Dracula was the former's guilt and revulsion at his own bloodlust: he was basically a good man in the grip of an uncontrollable addiction. Like other supernatural characters of the 1970s, Morbius was revived in his own series in the 1990s in Marvel's "Midnight Sons" line of titles, this time as a vigilante who, since he had to drink blood to survive, chose to prey upon criminals he judged unworthy of life.

Spirits of Vengeance #1 (1992) Art: Adam Kubert

Mistakenly thinking Dan Ketch was Zarathos, an older John Blaze hunted him down, intending to destroy him. Instead Blaze became an ally of the new Ghost Rider and a friend to Ketch, who eventually was revealed to be Blaze's younger brother, in the series *Spirits of Vengeance*. Soon, the former Ghost Rider was spun off into his own series, *Blaze*, in which he travels the country with his carnival, encountering strange manifestations of the occult wherever he goes.

Ghost Rider Vol. 2 #28 (1992) Script: Howard Mackie / Pencils: Andy Kubert / Inks: Joe Kubert

Although he looks identical to his predecessor, the second Ghost Rider is a supernatural being of unknown origin who became mystically linked with a teenager named Dan Ketch. The new Ghost Rider is driven by an overwheming compulsion to avenge the spilling of innocent blood. (Andy Kubert, who drew this splash page, and his brother Adam, represented by the *Spirits of Vengeance* cover above, are the sons of the great comics artist Joe Kubert, best known for his war comics, who inked this splash.)

"spilt innocent blood," and he chillingly reflected the violent rage characteristic of so many of today's vigilante heroes. His greatest weapon was the "penance stare" with which he tormented criminals, forcing them to suffer all the pain they have inflicted on their victims.

Ketch and Ghost Rider seem to be separate beings sharing the same human body, yet a survivor of an ancient race called the Blood, known only as Caretaker, has told Ketch that he and the Ghost Rider are actually two sides of the same being. Ketch refuses to accept the idea: the Ghost Rider regards himself as being on a necessary mission to combat evil, while Ketch is revulsed by Ghost Rider's violence and tries unsuccessfully to keep him from emerging into the real world. It is not unlike Bruce Banner trying in vain to imprison the Hulk.

Perhaps this explains *Ghost Rider*'s enduring popularity. The new *Ghost Rider* combines supernatural fantasy with the traditional Marvel themes of the super hero books, like teenage alienation and the divided self. The new Ghost Rider is also one of the new comics vigilantes driven by an intense hatred of today's criminal violence, a hatred that perhaps expresses the character's—and the readers'—own inner furies and demands to be released. That, too, is a traditional Marvel theme, whether it takes the form of the Hulk's rampages or Wolverine's berserker rages. The popularity of individual characters may wax and wane, and trends and fashions in comics may come and go, but the basic themes and archetypes remain, merely shifting their outward forms as the years and generations pass.

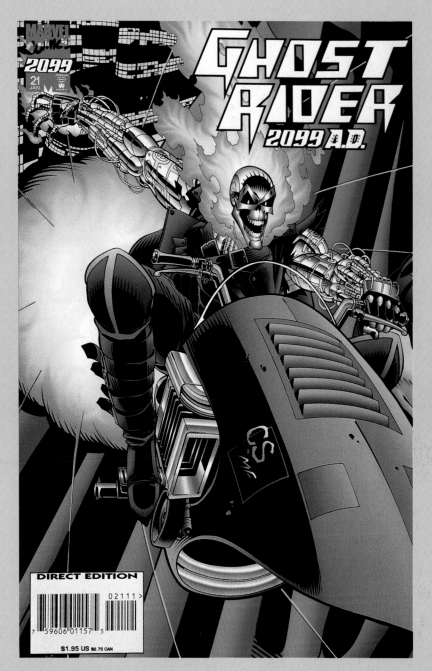

Ghost Rider 2099 #21 (1996) Pencils: Chris Sprouse / Inks: Mark Farmer

GHOST RIDER 2099

In the alternate future depicted in Marvel's 2099 titles, Zero Cochrane, a member of a street gang of hackers, survived his own murder. His consciousness entered cyberspace, where it encountered the entities making up the Ghostworks, computer networks that had achieved true sentience. Appalled at the way the major corporations are despoiling the planet, the Ghostworks entities resurrected Cochrane as their servant, creating a new cybernetic body for him that resembles a legendary skull-headed biker: thus Cochrane became Ghost Rider 2099, created by writer Len Kaminski and artist Chris Bachalo.

SIX

PROTECTORS OF THE UNIVERSE

The saga of Marvel's flagship team of super heroes, the Fantastic Four, began with a mission into outer space, and ever since Marvel has turned the entire universe into a stage for tales of heroic fantasy. After all, the modern Marvel line of super hero comics evolved from the company's "monster" and science-fiction books of the late 1950s and early 1960s, in which alien invaders from outer space were frequently featured. The second issues of both *Fantastic Four* and *Hulk* found the title characters in conflict with alien conquerors. Some Marvel super hero series, like *Spider-Man* and *Daredevil*, remained "grounded" on Earth, rarely even leaving New York, while other series, like *Fantastic Four* and *Thor*, ventured out into the cosmos from the very beginning. And still later, series like *The Silver Surfer* and *Warlock* made the break from Earth almost entirely.

GALACTIC EMPIRES

Uncanny X-Men #137 (1980) Script: Chris Claremont / Pencils: John Byrne / Inks: Terry Austin

Fantastic Four #91 (1969) Script: Stan Lee / Pencils: Jack Kirby / Inks: Joe Sinnott

There are three great galactic empires in the Marvel Universe: the Shi'ar, the Kree, and the Skrull. Above, Empress Lilandra of the Shi'ar (the second standing figure from the left) holds a summit via a futuristic form of video conference call with the Supreme Intelligence of the Kree (left), an entity created through the linkage of the preserved brains of great Kree of the past with immense computer systems, and the Skrull Empress Rk'lll (right), who later died in an attack on the Skrull Throneworld by Galactus and was succeeded by the Empress S'Byll. The shape-changing Skrulls especially provide opportunities for unusual story lines: in a tale (left) reminiscent of both *Spartacus* and a celebrated *Star Trek* episode, the Thing found himself enslaved and forced to fight as a gladiator on a Skrull planet whose inhabitants had adopted the forms and life-styles of 1920s Chicago gangsters.

Within the image:

NOT EVEN MY EXPERIENCES AS EARTH'S COSMIC DEFENDER HAS PREPARED ME FOR ANYTHING LIKE THIS.

I KNOW WHO THEY ARE FROM STORIES EON HAD TOLD ME.

BUT TO ACTUALLY BE IN THEIR COLLECTIVE PRESENCE IS OVERWHELMING.

THE MYSTERIOUS WATCHER!

THEY ARE THE ASTRAL DEITIES OF THE UNIVERSE!

ETERNITY, THE ACTUAL EMBODIMENT OF ALL THERE IS!

I WAS EXPECTING NEITHER ETERNITY NOR THE LIVING TRIBUNAL.

KRONOS, THE TITAN GOD OF TIME!

THEIR PRESENCE ALSO COMES AS A SURPRISE TO ME.

LORD CHAOS AND MASTER ORDER, THE GALACTIC BALANCE.

THE POWERFUL AND ENIGMATIC STRANGER!

THE MYSTERIOUS EMBODIMENTS OF LOVE AND HATE!

THE LIVING TRIBUNAL, THE COSMIC JUDGE OF ALL REALITIES!

THE MIGHTY DESTROYER OF WORLDS, GALACTUS!

AND TWO GIANTS THAT CAN ONLY BE CELESTIALS!

Infinity Gauntlet #3 (1991)
Script: Jim Starlin / Pencils: George Perez / Inks: Josef Rubinstein

The Earth-born hero Quasar, his patron Eon, the Silver Surfer, and Adam Warlock observe an unusual gathering of many of the "omnipotents" of the Marvel Universe. From left to right: Eternity; Uatu, the Watcher assigned to observe Earth; Kronos, an Eternal who rose to godhood; the conceptual being Lord Chaos; the Living Tribunal; Chaos's opposite, Master Order; the Stranger, who collects and experiments upon "lesser" beings; Galactus; two more conceptual beings, Mistress Love and Master Hate; and, in the lower right, two of the inscrutable Celestials.

AN ARRAY OF ALIENS

The term "Marvel Universe" is no misnomer. Other comics companies' stories in the 1950s and 1960s might deal with aliens, either friendly or malevolent, from individual fictional worlds. Marvel was different: it did not deal in single worlds but in galactic empires. Such races as the Skrulls and the Kree, the Colonizers of Rigel, and the Shi'ar would appear time and again in Marvel stories: not only do they intervene in the affairs of Earth, but Earth's own heroes become embroiled in their internal power struggles and their wars against each other. Most of the Skrulls and Kree were antagonistic to humanity, but extraterrestrials such as the Silver Surfer, the Rigellian Tana Nile from *Thor*, and the "living robot" Torgo from *Fantastic Four* became the friends and allies of Earth heroes; the X-Men's Professor Xavier even became for a time the consort of the alien princess Lilandra, empress of the Shi'ar.

If the Marvel Universe contains races that have emotions and goals similar to human beings, it is also home to the omnipotents, virtual immortals with agendas and purposes that can seem incomprehensible by our standards. They are nothing less than a pantheon of gods recast in secular, science-fiction terms. Among them are the Watchers and Galactus (see chapter one). Then there are the Elders of the Universe, the last survivors of now-extinct races, each of whom spends his or her endless existence in obsessive pursuit of a single field of interest, just as Ares was the god of war and Aphrodite the goddess of love.

The rivers, the sky, and the earth itself were gods in Greek mythology; so too entire heavenly bodies in the Marvel Universe prove to be both alive and sentient. Lee and Kirby dispatched Thor into the mysterious Black Galaxy to confront and defeat Ego the Living Planet. Doctor Strange and Captain Marvel both encountered Apalla the Sun Queen, a

THE LIVING UNIVERSE

Fantastic Four #234 (1981) Script and art: John Byrne

Among the sentient heavenly bodies in the Marvel Universe are Ego the Living Planet (left), created by Stan Lee and Jack Kirby, a being of nearly limitless power driven by a monstrous will to dominate everyone and everything, and a newborn nebula that descended from the heavens to Earth, took on human form, and joined the Defenders under the name of Cloud. Cloud (below, with Iceman) kept switching between two human forms, those of a teenage boy and girl, affording Defenders writer Peter B. Gillis the opportunity to explore her teammates' attitudes toward different genders.

Defenders #150 (1985) Script: Peter B. Gillis / Pencils: Don Perlin / Inks: Randy Emberlin

Silver Surfer Vol. 3 #4 (1987) Script: Steve Englehart / Pencils: Marshall Rogers / Inks: Josef Rubinstein

The first two Elders to appear turned up in *The Avengers*: the Collector, who attempted to add some of Earth's super-humans to his encyclopedic museum of the cosmos, and the Grandmaster, who could alter time and space in the course of gambling against an opponent. Since then numerous others have turned up, such as the Obliterator, who single-handedly wages war on other races; the Champion, who challenges super-humanly strong beings to battle to test his own strength; and the Gardener, who nurtures vegetable life throughout the universe. A conclave of many of the Elders of the Universe once met on the surface of the living planet Ego (left). From left to right are the Grandmaster; the Runner; the Collector; the Champion; the Trader; the Possessor, a seeker of knowledge; the Astronomer (standing in front of one of Ego's guards, an "antibody" in semihuman form); the Gardener; the Obliterator; and the Contemplator, originally named "Mister Buda" by his creator, Jack Kirby.

star manifesting itself as a humanoid being.

Higher than all of these in the hierarchy of the Marvel Universe stand the "abstract" and "conceptual" beings, entities that embody some aspect or governing principle of the cosmos, such as Death and Eternity, who together are said to constitute the entire fabric of the universe, and Jim Starlin's pairings of Master Order and Lord Chaos and Mistress Love and Master Hate. These deities most often turned up in the adventures of the sorcerer Doctor Strange, but over the last decade especially, they have increasingly appeared in Marvel's science-fiction series as well.

This universe of remarkable beings is explored by a colorful array of space-traveling heroes. Some of them, like Marvel Boy, Nova, and Quasar, are Earth-born human beings who were granted extraordinary abilities by benevolent aliens. But two of Marvel's most celebrated space adventurers, Captain Marvel and the Silver Surfer, were aliens themselves, cast into the role of heroes through their contact with the people of Earth.

THE FIRST BOY IN SPACE

Marvel Boy, Marvel's first major "cosmic" hero, debuted in his own comic in 1950, drawn by Russ Heath and Bill Everett and named after a Simon and Kirby creation of the previous decade. This new Marvel Boy was Bob Grayson, whose scientist father had taken him as an infant to live with the advanced humanoid race inhabiting the planet Uranus. (Comics writers in the 1950s never seemed to concern themselves with simple matters like the fact that Uranus's high gravity, low temperatures, and poisonous atmosphere made it utterly uninhabitable.) Still, the most "cosmic" aspect of this short-lived series was its hero's origin. The teenaged Marvel Boy returned to Earth as a costumed crime-fighter, employing the powers of the unusual bands he wore on his wrists, supposedly products of Uranian science.

The first modern-day "cosmic" Marvel hero to receive his own series, before even the Silver Surfer, was Captain Marvel, but not the Captain Marvel

MARVEL BOY

Marvel Boy's 1950s series lasted only two issues (below), but Roy Thomas remembered him and revived him in *The Fantastic Four* as an adult called the Crusader, recapping his origin in a flashback to his youth on Uranus and return to Earth (bottom). Driven mad by the deaths of the entire Uranian populace, the Crusader perished in combat against the Fantastic Four when his power bands overloaded with energy. His body was disintegrated, but the bands survived, with repercussions in a later series. (Later, a member of the New Warriors was known as Marvel Boy before changing his name to Justice.)

Marvel Boy #1 (1950)
Art: Russ Heath

Fantastic Four #165 (1975) Script: Roy Thomas / Pencils: George Perez / Inks: Joe Sinnott

many older readers of this book will think of first. The original Captain Marvel was not a Marvel character at all: he was Billy Batson, the young orphan who spoke the magic word "Shazam" in order to transform into "the world's mightiest mortal." Published by Fawcett starting in 1939, this Captain Marvel rivaled Superman in popularity until his adventures ended in 1951; eventually DC Comics acquired publication rights to this Captain Marvel in the 1970s.

But in 1968 the name "Captain Marvel" was up for grabs, and the company's lawyers believed Marvel Comics should grab it. So Stan Lee and artist Gene Colan came up with a new Captain Marvel in the pages of *Marvel Super Heroes*; Roy Thomas took over as writer when the Captain soon moved into his own comic. By the kind of coincidence that strains credibility, this hero was really named Mar-Vell, a captain in the space fleet of the Kree, an imperial alien race that had already clashed with the Fantastic Four. Mar-Vell was sent to Earth as a spy and adopted the identity of the late scientist Dr. Walter Lawson, who, by another unlikely turn of events, was Mar-Vell's exact double. As Lawson Mar-Vell infiltrated the NASA base at Cape Canaveral, where he had to deal with the suspicions of the young and beautiful security chief Carol Danvers. Inevitably, he ended up going into action in his Kree military uniform (and identity-concealing helmet) to combat menaces such as a reactivated Kree Sentry robot and the Super-Skrull; Earthmen, hearing his name, thought he was calling himself "Captain Marvel" and this name stuck. Having thus become their

THANOS

Thanos (left) literally worshiped Mistress Death, who is a sentient, "conceptual being" in the Marvel Universe and appeared to him as a hooded female figure. The mad Titan sought to achieve absolute power so he could please his "Mistress" by wiping out all life in the cosmos, yet she remained silent, impassive, forever unattainable. Originally Jim Starlin made Captain Mar-Vell Thanos's principal nemesis, but he finally had Adam Warlock finish Thanos off by turning him to living but immobile stone. In recent years Starlin revived Thanos in books such as *Silver Surfer* and the limited series *Infinity Gauntlet*, *Infinity War*, and *Infinity Crusade*, where he became the nemesis of other heroes. Thanos created an immense shrine to Mistress Death, including gigantic busts of both her human and skeletal guises. In the panel below, Thanos stands atop the staircase, with Mistress Death herself at its base, and the demon Mephisto to the left.

THANOS DEFENDS HIS DIVINITY.

LET THE HEAVENS TREMBLE.

▲ *Infinity Gauntlet* #5 (1991) Script: Jim Starlin / Pencils: Ron Lim / Inks: Josef Rubinstein

Infinity Gauntlet #1 (1991) Script: Jim Starlin / Pencils: George Perez / Inks: Josef Rubinstein with Tom Christopher

Captain Marvel (First Series) #17 (1969) Script: Roy Thomas / Pencils: Gil Kane / Inks: Dan Adkins

Mar-Vell could escape the Negative Zone and come to Earth by trading places with the youth Rick Jones, the former sidekick of the Hulk and Avengers, when Rick struck together the Kree "nega-bands" he wore on his wrists. (Hence, Thomas and Kane devised a parallel between Mar-Vell and Jones and the original Captain Marvel and the younger Billy Batson.)

Marvel Graphic Novel #1: The Death of Captain Marvel (1982) Script and art: Jim Starlin

The heroes of the Marvel Universe gathered to pay tribute to their colleague Mar-Vell as he lay on his deathbed. Leaning on the bed is Elysius, an Eternal from Saturn's moon Titan, who was Mar-Vell's lover and became the mother of his son, the current Captain Marvel.

Captain Marvel (Third Series) #2 (1996) Pencils: Ed Benes / Inks: Mike Sellers

Genis, alias Legacy, created by writer Ron Marz and artist Ron Lim, mourns the memory of his father, the Kree warrior Mar-Vell. His aging having been artificially accelerated, the adult Genis has adopted the name his father bore on Earth, "Captain Marvel."

champion, Mar-Vell found himself developing sympathy for the human race he was supposedly to regard as the enemy.

No other Marvel hero has gone through such radical revampings as Captain Marvel did through his history. Mar-Vell rebelled against both his spy mission and his field superior, Colonel Yon-Rogg; fleeing from Earth he encountered a godlike being called Zo, who granted him additional superhuman abilities so he could destroy Yon-Rogg. Subsequently, under writer Archie Goodwin, Zo proved to be an illusory construct devised by two high-ranking Kree officials in order to manipulate Mar-Vell into acting as their pawn in their plot to take over the Kree empire. ("Zo" is "Oz" backward, get it?) After defeating the conspirators, Mar-Vell, now in the hands of writer Roy Thomas and artist Gil Kane, became trapped in the other-dimensional Negative Zone.

The best-remembered *Captain Marvel* stories came later, during the stint of writer and artist Jim Starlin, who pitted the Captain against his archvillain Thanos, the evil near-immortal from Saturn's moon Titan. Still later, writer Doug Moench had the cosmic entity Eon select Mar-Vell to be "protector of the universe," with his especial role

to be guardian of Earth, whose growing numbers of super heroes were the first wave in an evolving race of superhumans.

Starlin's finest Captain Marvel story was the character's last, Marvel's first graphic novel, *The Death of Captain Marvel*. During one tale in past years, when Captain Marvel had battled a criminal called Nitro, the Kree warrior had been exposed to a nerve gas that was now revealed to have been carcinogenic. Thus this graphic novel was a somber farewell to a Marvel hero who was dying not by a super-villain's hand in combat but through a disease that claims millions of ordinary people, surrounded on his deathbed by friends, his super hero allies, and even past foes, all

come to do honor
to him. Especially in
recent years, death has become something of a joke
in super hero comics, since virtually every character,
hero or villain, who is killed off in a story eventually
turns up alive. But, so far, Captain Mar-Vell's death has
remained final, perhaps because comics writers
have so much respect for the power of his final story.

Instead, the Captain has had a succession of heirs.
The name Captain Marvel was inherited by Monica
Rambeau, an African-American policewoman from
New Orleans who gained the power to transform
herself into any form of electromagnetic energy, and
served a long term with the Avengers. Beginning in
1996 Mar-Vell's son, aptly named Legacy, adopted
his father's name as the star of a brand-new *Captain
Marvel* series.

THE SILVER SURFER'S HUMANITY

The Silver Surfer's appearances in *The Fantastic
Four* had won him tremendous popularity amid
Marvel's growing and outspoken fan following in
the 1960s. It was inevitable that in 1968 the Silver
Surfer moved into his own series, written by Stan
Lee and drawn by John Buscema, where he evolved
the persona that would characterize him for the
next decade and a half.

This entailed a radical shift in the Surfer's characterization: whereas Jack Kirby had clearly conceived of the Surfer as an alien who had to learn to
be human, Lee now revealed him to have originally

SILVER SURFER ORIGIN

When Galactus came to Zenn-La, intending to make it
his next meal, Norrin begged that his world be
spared. Impressed by the mortal's courage, Galactus
offered him a Faustian bargain: he would spare Zenn-
La if Norrin would agree to abandon the planet and
his beloved Shalla Bal and become his herald and scout
for other worlds to victimize. Norrin agreed (left) and
was transformed into the Silver Surfer (below). He
had been exalted to godhood at the price of helping
to doom other worlds. (Later writers explained that
Galactus used his powers to manipulate the Surfer's
mind, slowly corrupting him into the amoral destroyer
that Kirby had first introduced.)

Silver Surfer Vol. 1 #1 (1968)
Script: Stan Lee / Pencils: John
Buscema / Inks: Joe Sinnott

BUT, EVEN AS NAMELESS *FEAR* AND MOUNTING *PANIC* SPREAD O'ER THE EARTH LIKE SOME FOUL AND FATAL *FUNGUS,* WE TURN OUR ATTENTION *ELSEWHERE*--TO A WORLD *BEYOND* THE FARTHEST BORDER OF IMAGI- NATION--YET, *CLOSER* THAN THE NEAREST *NIGHTMARE--!*

IF WE *LISTEN*--WITH RAPT ATTENTION--WE SHALL HEAR A *VOICE--* LIKE NONE YOUR *HUMAN EAR* HAS EVER HEARD-- LIKE NONE THE *HUMAN MIND* CAN E'ER CONCEIVE--!

I SENSE A STRANGE *DISTURBANCE* UPON THE WORLD WHICH MEN CALL *EARTH!*

THE WORLD WHICH HAS EVER BEEN THE MOST ABUNDANT *SPAWNING GROUND* FROM WHICH I PEOPLE MY *DOMAIN!*

IF YON *BREEDING PLACE* NOW BE THREATENED-- *MEPHISTO* MUST KNOW OF IT!

WHILE MAN REMAINS AN EDUCATED *SAVAGE,* MY RANKS OF THE *DAMNED* ARE *SWELLED* TO OVER- FLOWING!

THUS, I HAVE ORDAINED THAT *NOTHING* SHALL CHANGE THE UNTHINKING MASSES OF HUMANITY!

FOR, SO LONG AS THEY *REMAIN* AS HOSTILE, AS CONSUMED WITH *GREED* AND *HATRED* AS THEY ARE--

SO LONG SHALL THE ULTIMATE *VICTORY* BE *MINE*--ON THE DAY OF *ARMAGEDDON!*

DO NOT BE *ALARMED!* I AM HIM WHO WAS...*NORRIN RADD!*

THEN...IT'S *TRUE!* YOU *REACHED* THE FLYING *GLOBE!* BUT, WHAT HAS *HAPPENED?* WHAT HAVE THEY *DONE* TO YOU?

MY FATE IS OF LITTLE *CONSEQUENCE!* SUFFICE IT TO SAY...THIS PLANET SHALL *NOT PERISH!* BUT LET NOT THE *SPIRIT* OF OUR ANCEST- ORS BE LOST A SECOND TIME! LET NOT OUR PEOPLE GROW SOFT AND INDOLENT!

HE WHO COMMANDS THE GLOBE WILL SOON *DEPART...* AND *ZENNA-LA* SHALL RISE AGAIN!

YOU... SOUND AS THOUGH...

YOU WILL *NO LONGER* ...BE *AMONG* US!

Siver Surfer #1 (1968)
Script: Stan Lee / Pencils: John Buscema / Inks: Joe Sinnott
Star-crossed lovers: Silver Surfer and Shalla Bal, soon to be separated by uncounted light years.

Silver Surfer #3 (1968)
Script: Stan Lee / Pencils: John Buscema / Inks: Joe Sinnott
At the top of the list of the angelic Silver Surfer's nemeses was the demon Mephisto, who has unceasingly sought to reduce the Surfer to despair and submission.

been Norrin Radd, a human from the paradisiacal planet Zenn-La, forever exiled from his home and his idyllic relationship with his lover, Shalla Bal, to become an unwilling herald of Galactus.

Condemned to live on Earth, longing for his freedom and, even more, for the long-lived Shalla Bal, who remained amazingly faithful to him after centuries of separation, the Surfer was enthralled by the beauty of his new world and equally appalled by man's devastation of the planet and his cruelty to his fellows. Dragged down from the heavens, he now found himself the victim of persecution by fearful humans and the target of evildoers. Through this series Stan Lee had, in effect, found a way to recast the medieval morality play, with the Surfer as the sinless knight, pledged to a true love he cannot have, suffering through temptation by the Devil himself. Lee attempted to give the Surfer stylized, elevated diction to express his noble sentiments, while Buscema would often show him writhing in physical agony over his own emotional torments. In truth, the whole effect could often go over the top, and the Surfer sometimes seemed more self-pitying and helpless than stalwart and noble. Perhaps this is why the Surfer's original series lasted a mere two years.

It was not until the 1980s that the Surfer starred in a commercially successful series, which continues into the present. Writer Steve Englehart finally freed the Surfer from his entrapment on Earth, and he and subsequent writers have usually kept him in outer space, intervening in intergalactic wars, combating the machinations of the death-worshiping villain Thanos, and trying to come to terms with a Shalla Bal who now has more in her life than waiting for him: namely, her own career, as queen of Zenn-La. This Surfer is more quick to anger, less averse to battle, and even more open to romances with other women than the Silver Surfer of the 1960s. Perhaps the brooding, philosophical Surfer of a quarter century ago is out of place in the more violent comics adventures of the present; then again, perhaps he is more needed now than ever.

THE MAN-MADE GOD: ADAM WARLOCK

After creating the Silver Surfer, Lee and Kirby portrayed another godlike being come to Earth: "Him," also introduced in *The Fantastic Four*. "Him" was an artificial man created on an island called the Citadel of Science to be the first of a master race to conquer the planet. As with the Surfer, the first living soul to reach out to "Him" was Alicia Masters, who seemed to have an intuitive rapport with aliens and monsters. When she found "Him," this superior being was still unborn, a fetus within an adult-sized "cocoon," but possessing a fully formed intelligence and the ability to talk, frightened of the unknown outside this artificial womb. Upon emerging as an adult, "Him" was revulsed by the greed and ambition of his creators and obliterated the Citadel.

Some years later writer Roy Thomas and artist Gil Kane collaborated on recasting "Him" as Adam Warlock, the central character of his own series. Another Lee and Kirby character, the High Evolutionary,

Secret Wars II #3 (1985) Pencils: Al Milgrom / Inks: Steve Leialoha

Infinity Crusade #1 (1993) Pencils: Ron Lim / Inks: Al Milgrom

EPICS

In the 1980s Marvel began issuing limited series that teamed most of its leading heroes in a cosmic setting. The first of these was writer Jim Shooter's *Marvel Super Heroes: Secret Wars*, in which an unseen entity called the Beyonder transported super heroes and super-villains to a planet of his own creation, the better to study them. Shooter's ambitious *Secret Wars II* explored the perennial mythic motif of the god who descends to Earth in human form. Unwilling to restrain such newfound human emotions as rage, the Beyonder spread disaster and had to be stopped. Jim Starlin continued the trend into the 1990s, pitting the Marvel heroes against Thanos in *The Infinity Gauntlet* and against their own evil alien doppelgängers in *The Infinity War*. In *The Infinity Crusade* Starlin again turned to Jungian motifs in creating the Goddess, Warlock's corrupted anima come to life, who sought universal domination under the guise of religion.

Fantastic Four #67 (1967) Script: Stan Lee / Pencils: Jack Kirby / Inks: Joe Sinnott

Above, the blind Alicia Masters discovers the womb-like cocoon in which the artificially created Adam Warlock awaited his birth. As yet nameless, the being emerged to confront the leaders of the Enclave, who foolishly thought they could make him their slave (above right).

The character of Adam Warlock was more fully developed in stories written and illustrated by Jim Starlin in the 1970s (right). Forsaking the Christ analogies of past story lines, Starlin turned to pitting Warlock against the leaders of repressive religions he invented for other worlds, and, even more interestingly, had him dabble in Jungian psychology. Aside from Thanos, Warlock's greatest nemesis in the 1970s was the Magus, his own evil "shadow" self, who came to physical life and carved out his own interstellar empire.

Strange Tales Vol. 1 #179 (1975) Script and art: Jim Starlin

created "Counter-Earth," a duplicate of Earth on the other side of the sun, populating it with a new human race innocent of evil. But one of the Evolutionary's previous creations, the lupine Man-Beast, playing Satan, introduced evil into this new world. Warlock, acting as "son" to the High Evolutionary, descended to Counter-Earth to save it from the Man-Beast, accumulated a band of apostles, suffered crucifixion, and inevitably rose from the dead. This series was for the most part as obvious as this summary suggests, but Warlock finally came into his own in the 1970s in the hands of writer/artist Jim Starlin. Although Starlin's original run on *Warlock* was relatively brief, he returned to the character in the 1990s with great commercial success in series such as *The Infinity Gauntlet* and *Warlock and the Infinity Watch*.

FROM EARTH TO THE STARS

If Captain Mar-Vell was a warrior from space who became a defender of Earth, then Nova and Quasar were ordinary Earthmen who rose above their mediocre beginnings to become defenders of the entire cosmos.

Writer Marv Wolfman and artist Sal Buscema created Nova in 1976 as that decade's counterpart to Spider-Man. Like Peter Parker in his original tales, Richard Rider was a teenager in high school; unlike Parker, he had living parents and a brother. Whereas Spider-Man was usually confined to New York City, Nova, though based in Long Island, eventually discovered his sphere of operations included

outer space. Rider was endowed with superstrength and the ability to fly by a dying Nova Centurion, one of the elite warriors of an alien humanoid race called the Xandarians. Calling himself Nova the Human Rocket, he began his career as a costumed crimefighter in New York, but eventually traveled into outer space, where he joined the Xandarians in combating their Skrull enemies.

Nova's original series proved short-lived and he disappeared from the Marvel Universe for years, but in 1990 he returned as a member of the New Warriors. Writer Fabian Nicieza took him back into outer space, where he joined a new team of Nova Centurions composed of members of various alien races banded together to combat perils to their worlds.

Like Richard Rider in *Nova*, Wendell Vaughn, also known as Quasar, was an Earthman of only middling ability in everyday life who had cosmic power and tremendous responsibilities thrust upon him by fate. (Once more, Spider-Man's theme that "With great power must come great responsibility"

Nova #19 (1978)
Script: Marv Wolfman /
Pencils: Carmine Infantino /
Inks: Tom Palmer

When an alien Nova Centurion warrior died, an unsuspecting Long Island teenager, Richard Rider, inherited his powers and became Nova the Human Rocket.

QUASAR

Twice in his series Quasar underwent literal death and resurrection, once in the course of the series' most ambitious story line. This saga pitted Quasar against an Earthman named Maelstrom, the servant of the conceptual being Oblivion, who may be another aspect of Starlin's Mistress Death. Maelstrom intended to sacrifice the entire universe to his master by plunging it into a colossal black hole. But first he slew Quasar, whose spirit then came into contact with Oblivion's opposite, Infinity, a benevolent female entity, who embodied the life forces of all living beings in the cosmos. Through her intercession, Quasar was able to descend into the black hole—a science fictional counterpart to a mythological realm of death—and defeat Maelstrom (below), saving the cosmos and enabling himself to return to physical life, which he did in a new costume (above), symbolizing his rebirth.

Quasar #25 (1991) Script: Mark Gruenwald / Pencils: Greg Capullo / Inks: Keith Williams

influences another Marvel series.) Not only is he the inheritor of the power bands, now known as quantum bands, that were once worn by the original Marvel Boy, but early in his own series he was chosen by the late Captain Marvel's patron Eon to become the new "Protector of the Universe." Written by Mark Gruenwald, the *Quasar* series showed how this seemingly ordinary man, who had long regarded himself as a loser, came to terms with his new-found abilities, responsibilities he could barely even begin to comprehend, and a universe filled with wonders, horrors, and complexities that were beyond his imaginings.

The most memorable of the major story lines in *Quasar*, however, dealt with religious themes recast in Marvel's own mythology of "conceptual beings" and astral planes. Although he was an atheist, Quasar found himself dealing with an array of virtually omnipotent alien beings who could easily lay claim to the name of gods. Even at the series' outset, Quasar found himself contending against the conceptual being Deathurge, the literal embodiment of his own suicidal despair over the past course of his life.

Despite its space operatic facade, *Quasar* was at its best a challenging and experimental fantasy series, and succeeded in running for five years, finally concluding in 1994. It was followed in late 1995 by a limited series, *Starmasters*, in which Quasar teamed with the Silver Surfer, Beta Ray Bill from *Thor*, and new alien characters to form a team of adventurers joining him in his mission to be "Protector of the Universe."

Eternals Vol. 1 #1 (1976) Script and pencils: Jack Kirby / Inks: John Verpoorten
Jack Kirby recounts the history of the three species of humanity: the demonlike Deviants deep within the earth, the godlike Eternals on the mountaintops, and the rest of humankind, striving to survive between these two realms.

ETERNAL MYSTERIES

Jack Kirby returned for his final period at Marvel in the mid-1970s and created, wrote, and drew a series that is to my mind his last great achievement at Marvel, *The Eternals*. This is yet another recasting of the themes that can be seen in his work on *The X-Men*, *Thor*, the Inhumans, the New Men, and his *New Gods* mythos for DC: the superhuman race divided into two factions, angels and devils recast in terms of science fiction, battling for supremacy amid ordinary men and women like ourselves.

Like much of Kirby's work for Marvel and DC in the 1960s and 1970s, *The Eternals* is an inquiry into the nature of God. Working with Lee, Kirby had created the Stranger (in *X-Men*), Odin and the High Evolutionary (in *Thor*), the Source (in *The New Gods*), the Watcher and Galactus (in *The Fantastic Four*); now, working on his own in *The Eternals*, he presented us with his "space gods," the Celestials.

This is a race of giants, a hundred feet in height, their features and bodies masked by alien armor, who never speak. Their true appearance, their origin, and their motives are all enigmas. Three times they had come to Earth—the series opened on the eve of their fourth appearance. The First Host conducted genetic experimentation on the primates that are man's ancestors. As a result, three sub-species of humanity would evolve. First, there was normal humanity (later writers would establish that the Celestials implanted the genetic material that would enable twentieth-century humans to develop

Eternals Vol. 1 #6 (1976) Script and pencils: Jack Kirby / Inks: Mike Royer
Anthropologist Dr. Samuel Holden reveals the existence of the Eternals and the Deviants to an audience at New York's City College. From left to right are the Eternal speedster Makkari; the Eternal princess Thena; her sometime lover, the Deviant Warlord Kro (who has temporarily used his powers to give himself devil-like horns); the pleasure-loving Eternal Sersi; the Eternals' human friend Margo Damian; and her future lover Ikaris, warrior against the Deviants.

MAMMOTH CONCEPTIONS

Jack Kirby filled *The Eternals* with stunning visualizations of his epic concepts. There was the immense Celestial named Arishem (above), standing atop his pylons, towering above Aztec ruins, as he inscrutably pondered whether or not to destroy the human race at the end of fifty years. Eson (opposite), another Celestial, strode into the Deviants' subterranean kingdom of Lemuria, breaking down the walls that separated it from the seas above, unleashing a flood that swept through their city.

Eternals Vol. 1 #9 (1977) Script and pencils: Jack Kirby / Inks: Mike Royer

Eternals Vol. 1 #3 (1976)
Script and pencils: Jack Kirby /
Inks: John Verpoorten

super powers). Then, there were the Eternals, perfect specimens of humanity possessed of super-human powers, virtually immortal and indestructible, who lived on mountaintops and were mistaken by ancient civilizations for gods. Finally, there were the Deviants, the Eternals' perennial adversaries. These were grotesque creatures with unstable genetic structures, so that no two looked anything alike, whose civilization was driven by a lust for power.

Millennia ago, the Deviants came close to dominating the entire world, reducing humanity to slavery; however, they over-reached themselves when they attacked the approaching starship of the Celestials' Second Host. The Celestials unleashed a weapon, apparently nuclear, which sank the Deviants' island kingdom of Lemuria under a vast flood. An ark brought one human family and its animals to safety, guided by the Eternal named Ikaris, who looked from the ship as if he were a dove.

A thousand years ago the Celestials' Third Host arrived and established a base amid the civilization of the Incas in South America. As *The Eternals* opened, Ikaris, in human guise, guided a human archaeologist and his daughter to this site, where they witnessed the arrival of the Fourth Host.

POWER PACK

Power Pack was a series unlike anything else at Marvel before or since: the super heroes in this book were all children, the youngest of whom was only five years old. The original stories, by writer Louise Simonson and artist June Brigman, created a wonderful atmosphere of gentle, childlike fantasy, transposed from the world of fairy tales to science fiction. Alex, Julie, Jack, and Katie Power were the children of the scientist Dr. James Power, who had created an invention that was sought by evil reptilian aliens called the Snarks (named after a character created by Lewis Carroll). The benevolent alien Kymellian, Aelfyre Whitemane, who resembled a humanoid horse, traveled to Earth in his sentient, talking starship to thwart the Snarks but failed. Dying, he transferred each of his superhuman abilities into one of the Power children, who then had to save their parents from the Snarks' clutches. Though the initial story line and others took the Power children into outer space, the series was primarily set in the more realistic locale of Manhattan's Upper West Side. Indeed, one of the series' strengths was its successful blend of Power Pack's fantastic adventures with a realistic depiction of the children's personalities and everyday lives. *Power Pack* lasted for several years and lives on through the eldest of the Power kids, Alex, who has since graduated to membership in the teenage New Warriors.

Eternals Vol. 2 #2 (1985) Script: Peter B. Gillis / Pencils: Sal Buscema / Inks: Al Gordon

Unlike so many blustering Marvel villains, the Deviant high priest Ghaur exuded a sinister, commanding authority through his regal stillness and the subtle threats in his carefully chosen words. Yet this quiet menace possessed the greatest case of hubris imaginable: he sought to seize the power of the Dreaming Celestial, a renegade whom the other Celestials had buried beneath a mountain range, and thus make himself one of the "space gods."

People were awestruck and frightened when these alien giants suddenly appeared at sites all over the globe, and then the Deviants chose to emerge from their undersea lair and spread chaos amid the human race. Ikaris, especially, set himself to combat the Deviant menace to humanity.

The Eternals is as memorable for its characters as it is for Kirby's epic feats of visualization. There was the shadowy, brooding figure of the Forgotten One, the Eternal who was known to ancient civilizations as Gilgamesh, Samson, and Hercules. (Kirby seemed untroubled by the fact that he had already co-created a Hercules in *Thor*; Roy Thomas eventually had Hercules and the Forgotten One meet.) There was Kro, the demonic military leader of the Deviants, who despite his ruthlessness was still

Power Pack #3 (1984) Pencils: June Brigman / Inks: Bob Wiacek

KILLRAVEN

Marvel Graphic Novel #7: *Killraven* (1983) Script: Don McGregor / Art: P. Craig Russell

Roy Thomas came up with the idea of doing a comic book sequel to H. G. Wells's classic science-fiction novel *The War of the Worlds,* the saga of an unsuccessful invasion of Earth by Mars in 1901. Marvel's continuation began in *Amazing Adventures* in 1973 with a story written by Gerry Conway and illustrated by Neal Adams, showing how the Martians returned in 2001 and this time swiftly conquered the planet. Years later, Killraven, a youth raised to fight in the Martians' gladiatorial arenas, escaped and organized a small band of freedom fighters, sworn to free Earth and take the war back to Mars itself. (Killraven, by the way, exists on an "alternate time line" in which most Marvel super heroes were slain by the Martian invaders; it is not the "real" future that awaits Spider-Man and his contemporaries.) It was writer Don McGregor who transformed the *Killraven* saga, as it was renamed, into a classic. Of all of Marvel's writers, McGregor has the most romantic view of heroism. Killraven and his warrior band were also a community of friends and lovers motivated by a poetic vision of freedom and of humanity's potential greatness. McGregor's finest artistic collaborator on the series was P. Craig Russell, whose sensitive, elaborate artwork, evocative of Art Nouveau illustration, gave the landscapes of Killraven's America a nostalgic, pastoral feel, and the Martian architecture the look of futuristic castles.

Below, Cape Canaveral in the year 2020, converted by the Martians into their own scientific base.

Guardians of the Galaxy #1 (1990) Pencils: Jim Valentino / Inks: Steve Montano

GUARDIANS OF THE GALAXY

Killraven was not Marvel's first saga of freedom fighters battling alien conquerors. Back in *Marvel Super Heroes* #18 (1969), writer Arnold Drake and penciller Gene Colan had created the Guardians of the Galaxy, set in the thirty-first century of an "alternate future." In this era humans had colonized other planets in the solar system and even Alpha Centauri's, but Earth's entire interplanetary empire fell before the Badoon, a warlike race of sentient reptiles. They found their nemesis in Vance Astro, an astronaut from the twentieth century, who, inspired by the memory of his childhood hero, Captain America, formed a small band of freedom fighters to liberate Earth. The Guardians disappeared from sight after only one story, but resurfaced in the 1970s in a short-lived series written by Steve Gerber. Their most successful vehicle, originally written and illustrated by Jim Valentino, debuted in 1990 and ran for five years.

Along the top of this cover, from left to right, are Charlie-27, a human who had genetic alterations that gave him enough superhuman strength to live under Jupiter's heavy gravity; the mysterious Starhawk, who, caught in a time loop, repeatedly relives his life; and his foster sister and former wife, Aleta, who manipulates light energy. Along the bottom, from left to right, are Martinex, a human altered to live in Pluto's frigid climate; Nikki, who was genetically made capable of living in Mercury's heat; Yondu, an alien from one of Alpha Centauri's planets; and Vance Astro. (Astro is the counterpart on the Guardians' timeline of Vance Astrovik, the New Warrior known as Justice.)

gripped by passion for his former lover, Thena, the fiery warrior daughter of Zuras, monarch of the Eternals. And there was Sersi, perhaps the most fascinating of all, an Eternal with many sides to her personality. She was known to the Deviants as Sersi the Terrible for her temper and her ability to alter the shapes of persons or objects at will, as when she transformed Ulysses' men to pigs in ancient times. (Sersi explained that Homer had misspelled her name in *The Odyssey*.)

Despite its considerable merits, the original *Eternals* series was not a commercial success, perhaps because Kirby dealt with his large cast of characters as a true ensemble, continually shifting the focus from one group in one issue to another set in the next; there was no central heroic figure who appeared in every story line. The next attempt to give the Eternals their own series was a twelve-issue run in the 1980s written first by Peter B. Gillis and later by Walter Simonson and illustrated by Sal Buscema. They introduced something missing from the Kirby original: a villain of grand stature. This was Ghaur (pronounced "gore," of course), the high priest of the Deviants, whose best scenes showed him using subtle intrigue to achieve his ends, in contrast to the melodramatic bluster of most Marvel villains.

The Eternals still did not win a sufficiently large readership, and so they have had to make do with guest appearances and the occasional special, although Sersi uncharacteristically turned full-time adventurer and joined the Avengers for several years.

HE **LOOKS** LIKE A MAN...HE **THINKS** LIKE A MAN...BUT NOWHERE IN THIS WORLD IS THERE **ANYONE** AS EXCITING AND DIFFERENT AS...

MACHINE MAN

OH, PLEASE HELP US! FREDDY **SLIPPED** OVER THE EDGE OF THIS CLIFF AND--!

THE SITUATION IS APPARENT! **CARELESS** HIKERS BREED THEIR **OWN** MISFORTUNE!

LOOK! HIS ARM IS EXTENDING-- LIKE A **LADDER!**

STAND BACK! I'LL DO WHAT I CAN FOR HIM!

Machine Man #1 (1978) Script and pencils: Jack Kirby / Inks: Mike Royer

The *Eternals* was Jack Kirby's finest work for Marvel during his brief return in the 1970s, but it was hardly the only outlet for his inexhaustible imagination. In addition to his new work on *Captain America*, Kirby also wrote and drew a new series based on Stanley Kubrick and Arthur C. Clarke's landmark 1968 film *2001: A Space Odyssey*. This comic eventually mutated into a vehicle for a new Kirby creation, Machine Man (left). Like the original Human Torch and the Vision, he was an artificial man with free will and the capacity for emotions. But whereas they looked more or less human and were acclaimed as heroes, Machine Man was clearly mechanical and became an outcast. A science-fictional adult version of Pinocchio, Machine Man nonetheless longed to be treated as human. Through *Machine Man* Kirby gave perhaps Marvel's most intriguing twist on the theme of the masked identity. Machine Man regarded the faceplate with humanlike features he wore as the symbol of the humanity he sought; without it, Machine Man felt empty. Yet the society that persecuted him, threatening to reduce him to the status of government property, possessed far less sensitivity and true humanity than he.

Kirby's oddest contribution to 1970s Marvel was *Devil Dinosaur*, the tale of a tyrannosaur of unusual intelligence, his hide burned red by intense heat, who befriended Moon-Boy, a furry protohuman (below). Like so many other pop culture tales of prehistory, from *Alley Oop* through *The Flintstones*, *Devil Dinosaur* fell prey to the mistake of having dinosaurs coexist with early man, when they were actually separated by tens of millions of years. The 1980s limited series *Fallen Angels* corrected the problem by establishing that Devil Dinosaur and Moon-Boy actually lived in the present on a far-off planet and then transporting them to a Manhattan warehouse. And New Yorkers think those legendary alligators in the sewers are a problem.

DEVIL WAS NOT ALWAYS THE UNDISPUTED MASTER OF THE VALLEY OF FLAME. BUT THE FEROCITY AND STRENGTH OF HIS KIND HAS CONTINUALLY OVERCOME THE COMPETITION FOR TERRITORY. THE ONE STRANGE ADDITION TO DEVIL'S LIFE-LIKE EXPERIENCE IS MOON-BOY... A COMPANION WHOSE EXISTENCE IS FIRMLY BOUND TO HIS GIANT PROTECTOR...

ONLY YOU MUST RULE HERE! ATTACK HIM-- NOW!!

Devil Dinosaur #1 (1978) Script and pencils: Jack Kirby / Inks: Mike Royer

SEVEN

VIGILANTES AND LAWMEN

Not every Marvel hero can smash through walls or fly. The heroes in this chapter depend on their sharp wits, athletic skills, and whatever weapons they choose to carry to overcome their adversaries. Some have relatively limited super powers: Daredevil, for instance, can hear someone whispering a block away, but he has no more strength or agility than a normal man could achieve through training. Others, like Shang-Chi, have no superhuman abilities whatsoever. Some, like Nick Fury, are official lawmen. Others, like the Punisher, operate outside the law to fight crime. All of them, even those who work with organizations, are temperamentally loners, challenging overwhelming odds on their own.

DAREDEVIL ORIGIN

In the early 1960s Marvel stories often tried to devise rationales for the heroes' costumes. In his first appearance (left), Daredevil, the son of a fighter, wears a costume reminiscent of a masked wrestler. In Matt Murdocks's case, however, the identification with the father also had a psychological basis. Below left, the relationship is established: boxer "Battling Jack" Murdock urges his son to devote himself to his studies. Like many American parents born before World War II, Jack dreamed that his son would lead a better, more prosperous life than he had. The dream, in fact, originated with Matt's mother, and Jack had promised her it would come true, as indeed it did, but Matt also needed to be like his father.

Matt's transformation into a "daredevil" occurred when he nobly rescued an aged blind man from certain death, only to be blinded himself. Just as Odin, in the Norse myths, gained wisdom by sacrificing an eye, Matt Murdock exchanged his normal sight for a superhuman means of visualizing the world around him. The panels below, showing the accident, come from a 1980 retelling of the origin story.

Daredevil #1 (1964) Script: Stan Lee / Art: Bill Everett

Daredevil #1 (1964) Script: Stan Lee / Art: Bill Everett

Daredevil #164 (1980) Script: Roger McKenzie / Pencils: Frank Miller / Inks: Klaus Janson

DAREDEVIL—IN THE BEGINNING

In 1964, Stan Lee and artist Bill Everett created the first and foremost of Marvel's super heroes without super powers that can be used for combat: Daredevil, "The Man Without Fear." Actually, Daredevil does have super powers of a sort: he is a blind man whose other senses have been heightened to superhuman acuity. These abilities enable him to sense the whereabouts of his opponents, but apart from them he is an ordinary man, no stronger or more agile than a real-life athlete could make himself.

Neither the name "Daredevil" for a costumed hero nor even the idea of a blind super hero was new: Lev Gleason Publications did a character called Daredevil in the 1940s, and DC's 1940s hero Doctor Mid-Nite was a blind man who used "infrared lenses" to see at night. As always, what distinguishes the classic Marvel heroes is not their outer trappings but their unique characterizations.

Daredevil provides yet another of the super hero genre's variations on the theme of the divided self. Matthew Murdock was the son of a prizefighter named "Battling Jack" Murdock, a man who claimed to be a widower (although later stories would indicate he was lying). The Murdocks lived in Manhattan's inner-city neighborhood of Hell's Kitchen. Matt idolized his father and wanted to be an athlete too, but Jack saw himself as a loser and pressured the child to focus all his energy on his schoolwork, as his mother would have wished.

A brilliant student with no life beyond his schoolbooks, he suffered the taunts of neighborhood boys, who mockingly labeled him "Daredevil." Enraged, he began leading a dual life, secretly devoting as much effort to training himself as an athlete as he did to his studies. Thus, in this case, leading a dual life actually seemed to make Matt a more rounded person, uniting the cerebral, associated with his mother, with the physical, associated with his father, and enabling him to channel his potential for violence into self-improvement.

Matt's plans for himself were nearly wrecked when one day he saw an old, blind man, dressed ominously in black, walk out in front of an oncoming truck transporting radioactive wastes. Matt hurled himself into the truck's path, knocking the blind man out of the way, only to be struck down by the vehicle himself; worse, a canister fell from the truck onto Matt's face, releasing a heavy dose of radiation, that early 1960s all-purpose medium of change.

The spooky accident permanently blinded Matt but more than compensated for what it had taken away by heightening his other senses. It even gave him a "radar" sense that enabled him to perceive persons and objects around him. Even though he could not see, he now had higher awareness of the world around him than any sighted person. Matt survived this symbolic death and rebirth to pursue his paths in life even more intensely, entering college and stepping up his secret self-training.

Desperate to finance Matt's college career, Jack Murdock, now far past his prime, had signed with a criminal promoter called the Fixer. Ordered to take a

dive in his next fight, Murdock, knowing Matt was in the audience, instead miraculously overcame his younger opponent in "maybe my last chance to do something to make my son *proud* of me!" In retaliation the Fixer had Jack Murdock gunned down on the street.

Despite his grief, Matt Murdock did not let his father's death stop him any more than his blindness had. He claimed his position as full-fledged adult in both his lives. First he graduated law school as class valedictorian and the very next day opened his own law office in partnership with his classmate Franklin "Foggy" Nelson, who has remained Matt's Sancho Panza, the portly, comical, loyal companion to the noble hero, in the series ever since. (Matt also met their new secretary, Karen Page, with whom he began the standard early 1960s Marvel romantic subplot: the hero and his female workmate are deeply in love with each other but too shy and insecure to admit it to each other.) At the same time his "secret" life also reached its turning point. He realized he could not pursue his life as a lawyer—the public, cerebral path—until his father's killers had been brought to justice. "But years ago I promised Dad that Matt Murdock would use his *head* . . . never become a

fighter, never depend on his strength, the way *Dad* did!" As far as this story is concerned, this promise has the force of a mythic vow that Matt cannot possibly break. So, like the good lawyer he is, Matt finds a loophole: "I'll see to it that Matt Murdock never *does* resort to force . . . but someone else will! . . . Somebody totally *different* from Matt Murdock! . . ." So he made a costume for himself and adopted as his *nom de guerre* the name the neighborhood bullies called him, "Daredevil," thus mastering the old insult by turning it into a proud boast. Since his new alternate identity served to fulfill his forbidden inner needs, it seems appropriate that Matt gave his mask devil's horns. As later *Daredevil* writer Frank Miller noted, Matt Murdock was a good Irish Catholic boy who chose as an adult to take on the outward appearance of the devil. Daredevil is Matt's shadow self who has turned an evil—violence—to serve society's need for justice.

Another paradox lay in the fact that while Matt Murdock was a lawyer working for justice within the legal system, Daredevil was a vigilante who stepped outside the framework of law to achieve justice. Stan Lee repeatedly showed Daredevil bringing a costumed criminal to justice and then defending him in court as Matt Murdock. This was certainly a charitable attitude for Murdock and his writers to take, respecting the rights even of the guilty, and stands in sharp contrast to the merciless mind-set of the Punisher and subsequent characters of his ilk.

The very next day after first taking on his masked identity, Daredevil confronted the men responsible for his father's death: he captured Slade, the actual gunman, and the Fixer died of a heart attack while fleeing this costumed figure of nemesis. With his father avenged, Matt/Daredevil could begin his new dual life in earnest.

There was no rationale given for Matt retaining his Daredevil identity after the Fixer's demise. In the 1960s it was generally accepted that anyone who had super powers would don a costume and either fight or perpetrate crime. It would take a reworking of Daredevil's origins published thirty years later to give Matt the motivation for continuing his career as a costumed crimefighter. Still, surely it would have seemed emotionally wrong, once Matt had awakened this long-suppressed side of himself, to bury the part of him that was Daredevil.

THE MERRY SWASHBUCKLER

Bill Everett left *Daredevil* after the origin story, and the tone of the series began to change. The origin took place in a realistic milieu, and the Fixer and his crew were character types from conventional crime fiction; Daredevil was the lone fantasy figure in this story, but his original black-and-yellow costume was not so far removed from a practical set of acrobat's tights, in contrast with more fanciful super hero costumes like Spider-Man's or Thor's. With issue #2, however, drawn by Joe Orlando, Daredevil met the Thing and battled his first super-villain, Spider-Man's antagonist Electro. The series had thus already settled into being a typical Marvel super hero series, reminiscent of Spider-Man, that other costumed crimefighting acrobat, but without distinguishing qualities of its own. Exceptions were several early issues illustrated by the great Wally Wood, who had already made a legendary reputation with his work on the EC comics of the 1950s. Among these was the story of the hopelessly outmatched Daredevil's brave attempt to stop another of the Sub-Mariner's rampages through Manhattan. Wood and Lee designed Daredevil's all-red costume (an appropriate color choice for a "devil"), and Wood created the visuals for one of Daredevil's strangest early foes, the Stilt-Man, an armored menace who strode through the city on immense, telescoping legs.

In retrospect, these two stories can be seen as pointing the way for Daredevil's later direction. The series hit its stride when John Romita, Sr., took over drawing the series and went full speed ahead once he, in turn, was succeeded by Gene Colan. Their work combined a handsomeness honed in their years on romance comics with a palpable flair for dynamic action scenes; it seemed to inspire Stan Lee, who now turned Daredevil into a modern-day swashbuckler, Douglas Fairbanks Sr., or Errol Flynn as 1960s super hero. Yes, Matt Murdock suffered

CAREFREE DAREDEVIL

Daredevil #22 (1966) Script: Stan Lee / Pencils: Gene Colan / Inks: Frank Giacoia and Dick Ayers

Daredevil #27 (1967) Script: Stan Lee / Pencils: Gene Colan / Inks: Frank Giacoia

Through the artwork of John Romita, Sr., and Gene Colan, Daredevil became an icon of sheer grace, swinging effortlessly on his trademark grappling hook (the crook of his blind man's cane) and line from one skyscraper to the next, far above the ordinary people on the sidewalks below. Apart from his superhuman senses, Daredevil was an ordinary man, yet Lee and Colan kept placing him in perils that seemed impossible to survive. The Beetle hurls him down a waterfall! Daredevil and the Trapster struggle atop a small flying platform hovering far above the streets below! Daredevil fights the Jester atop the Statue of Liberty! Daredevil temporarily loses his superhuman senses, thereby becoming truly blind, and nevertheless has to take on Mister Hyde and the Cobra, a team formidable enough to challenge Thor!

Not only did Daredevil always emerge triumphant, thanks to his miraculous acrobatic skills, but throughout this period he remained Marvel's premier master of comic banter. Even Spider-Man could seem morose in comparison to Daredevil swinging lightheartedly through the city and in and out of fights.

Daredevil #25 (1967) Script: Stan Lee / Pencils: Gene Colan / Inks: Frank Giacoia
Daredevil indulged his sense of humor to the utmost through his third identity, that of Matt's over-the-top hipster twin brother Mike, to the befuddlement of Matt's law partner, Foggy Nelson, and the delight of their secretary, Karen Page.

Poster. Pencils: Frank Miller / Inks: Klaus Janson
Through both his writing and artwork, Frank Miller not only revolutionized *Daredevil*, but also inaugurated the long line of "grim and gritty" comic book series that continues to this day.

from the obligatory Marvel angst, primarily in his love life, but in costume he provided the wittiest banter of any Marvel hero. He was indeed a daredevil, and he clearly loved it: his costumed identity was not so much a means to vent his inner violence as it was a release from the straitjacket of a nine-to-five existence.

However dangerous his adventures became in the 1960s, Daredevil retained an inexhaustible sense of humor and an enviable joie de vivre. Indeed, when Foggy and Karen stumbled upon his dual identity, Matt created a third identity to throw them off the track: he convinced them that Daredevil was really his twin brother, Mike Murdock, and clearly had the time of his life playing the part: loudly dressed, flirting with Karen, making wisecracks at Foggy's expense, telling corny jokes while projecting the attitude of someone who is convinced he is the essence of cool, and somehow being charming despite it all. Mike was clearly as much a release for the studious, dependable, somewhat dull Matt as Daredevil was.

THE MILLER DAREDEVIL

When Lee departed *Daredevil* in 1969 the wildly escapist aspects of the series left with him, and for many years thereafter it returned to conventional super hero melodrama. Not until 1981 did *Daredevil* undergo revitalization with the beginning of Frank Miller's run as series writer. (He had already been drawing it since 1979, collaborating with writer Roger McKenzie.) Indeed, Miller's first issue as writer, #165, marked the beginning of something very new in mainstream comics. His *Daredevil* is superficially regarded as the start of "grim and gritty" comics, stories portraying the world of the super heroes as a far more dangerous, less stable place, and the characters, even the heroic ones, as darker, more violent, more psychologically flawed than the saintly protagonists of old. In the lesser hands of numerous imitators of Miller and other leading comics writers of the last fifteen years, such stories become cynical, almost unremittingly bleak, and sometimes revel in their characters' own violence and amorality. Miller's best work is far superior to this. In effect he was carrying forward the revolution that Lee and his collaborators had begun in the 1960s: they gave their stories more psychological and dramatic force by depicting a more realistic world than the simpler tales of past comics and by giving the readers characters with recognizable flaws. Inevitably, with the

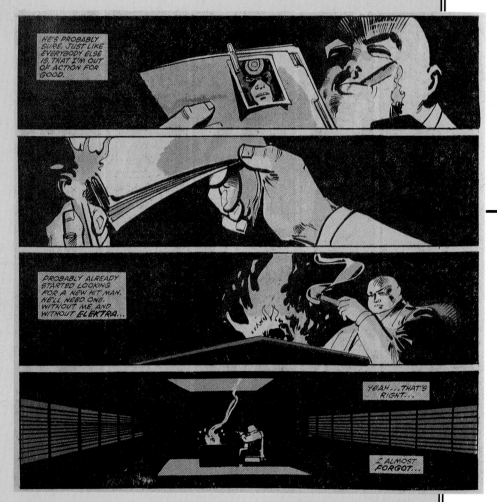

Caption text in panels:
HE'S PROBABLY SURE, JUST LIKE EVERYBODY ELSE IS, THAT I'M OUT OF ACTION FOR GOOD.

PROBABLY ALREADY STARTED LOOKING FOR A NEW HIT MAN. HE'LL NEED ONE. WITHOUT ME AND WITHOUT *ELEKTRA...*

YEAH...THAT'S RIGHT...

I ALMOST FORGOT...

Daredevil #181 (1982) Script and art: Frank Miller / Finished art: Klaus Janson

KINGPIN

The principal villain of Daredevil's world was a creation of Lee and Romita borrowed from *Spider-Man*, the Kingpin, who, until then, had seemed almost as unrealistic as Doctor Octopus: he was a blustering, virtually spherically shaped crimelord, who hurled himself into hand-to-hand combat with Spider-Man as if he were yet another bad guy in tights. Miller changed all of this: the Kingpin was still physically powerful, and was often shown engaging in workouts, easily overpowering his sparring partners, but more often now he could be found seated in his darkened office, his great bulk resembling an impassable barrier. From there he surveyed his empire, New York City, which he controlled through a ubiquitous web of criminal operations.

passage of time, in the hands of later writers, Marvel's 1960s innovations had themselves become conventions; the older readership Marvel had built for super hero comics was ready to move further. Miller gave Daredevil and a large cast of supporting players greater psychological complexity, thereby forcing them to struggle against their own flaws and passions as well as external adversaries. Miller reconceived the entire series—its tone, its characterizations, its worldview—while simultaneously molding it into a powerful expression of his own artistic personality. In doing so, he led the way for the numerous other revampings of classic characters in mainstream comics, at Marvel, DC, and elsewhere, that occurred from the mid-1980s onward.

Whereas most Marvel writers seemed content with doing variations on Stan Lee's own treatments of his characters, Miller reenergized Daredevil by calling upon new visual and narrative influences. The most important of these were crime fiction in the "hard-boiled" style of writers like Jim Thompson, a particular Miller favorite, and its cinematic counterpart, *film noir*. The latter was a film genre that originated in the 1940s with movies like John Huston's *The Maltese Falcon* and Howard Hawks's *The Big Sleep*. These books and movies created a world of troubled, morally flawed, but determined heroes moving through a shadow-laden world of powerful criminals and beautiful but treacherous women.

Daredevil #187 (1982) Script and art: Frank Miller / Finished art: Klaus Janson

STICK

One influence on Miller's *Daredevil* was his fascination with certain aspects of Japanese culture, notably the martial arts. Daredevil's mentor, Stick, was a blind old man with ninja skills who had developed his own radar sense without the benefit of Matt's radioactive accident. Stick was Matt's surrogate father in his secret life as a youth, the one who trained him in the use of both his superhuman senses and his fighting abilities. He served the role of *sensei*, but a particularly American version: a cranky old man who hung out in Manhattan pool halls, forever verbally abusing his pupil and even hitting him with his long stick when he performed below Stick's standards.

Hearkening back to Lee and Everett's origin tale, Miller returned Daredevil to a more realistic criminal milieu, in which costumed figures were few and thus had more startling dramatic power when they appeared. The super hero myth sets the lone crusader for justice not on a frontier but within a crowded, urban landscape. Lee and Ditko had conceived of organized crime in *Spider-Man* as a mélange of gangs out of the 1920s warring for control of the city's underworld; Miller depicted organized crime as a monolithic power that already held the city in its grip. The Kingpin, Daredevil's main nemesis, was no obvious thug, but the respected head of a corporation; he was not just one boss among many but the modern-day, uncrowned king of the city's dark side, who now reached out through his own mayoral candidate to dominate the light side as well.

Miller also gave the Kingpin another identity, his real name, Wilson Fisk. In Miller's third issue as writer, Fisk, on the wishes of his wife Vanessa, has forsaken his role of "Kingpin of Crime" and moved to Japan, becoming merely the "humble dealer in spices" that he publicly claimed to be. Fearing Fisk will use his records from his days as their leader to incriminate them, his former lieutenants kidnap Vanessa to force him to give them the files; in the ensuing conflict, Vanessa is seemingly killed by one of the men, Lynch, in an explosion. Believing "one moment of joy, my one brief instant of humanity" forever vanished, Fisk brutally murders Lynch and becomes a ruthless crimelord again, resuming control of his empire and setting the stage for his clashes with Daredevil, who regards himself as the city's guardian. But first he turns over the mob's files to Daredevil to use against his treacherous lieutenants. Daredevil will see justice done; the Kingpin rids himself of obstacles to his return to power. Daredevil accepts the files, aware that in doing so he has morally compromised himself to bring about what he hopes is a greater good; the choices in Daredevil's world are no longer clearly black and white.

There were only two other costumed figures in the world of Miller's Daredevil: his shadow and his

Daredevil #179 (1982)
Pencils: Frank Miller /
Inks: Klaus Janson

Elektra wields her trade-
mark weapon: the *sai*.

MARVEL COMICS GROUP

WIN A *Columbia* TEN-SPEED FORMULA 10. RACER!

DETAILS INSIDE

60¢ 179 FEB 02459

DAREDEVIL

...SOMEBODY HAD TO WIN!

anima. The former was Bullseye, who had been introduced by Marv Wolfman and Bob Brown into *Daredevil* back in 1976, an assassin with a flaw-less aiming ability, who can turn anything he could throw, even a paper airplane, into a lethal weapon. Miller transformed the char-acter into Marvel's first literally psychotic killer, the embodiment of evil as madness, a man who killed for sheer pleasure and was obsessed with destroying Daredevil, his nemesis and opposite. Bullseye was hired by the Kingpin to serve as his chief assassin, symbol of the viciousness through which he built his dark reign over the city. The other costumed figure was Elektra, a startlingly dif-ferent character for Marvel. She owed her role in *Daredevil* to Miller's continuing fasci-nation with hard-boiled detective fiction: she was the series' femme fatale, both the woman Matt Murdock loved and Daredevil's mortal foe.

The finale of Miller's original run on *Daredevil* took an extraordinary form: Daredevil playing Russian roulette with the unwilling, paralyzed Bullseye. In the course of the story Daredevil recalls an episode from his child-hood that explains one of the basic themes of Miller's work on this series. Here Miller gives a different slant to Matt's reaction to his father's insistence that he refrain from becoming a fighter. In Lee's version Matt was suppressing a natural athleticism; in Miller's version (here and reworked in his later

Daredevil #181 (1982)
Script and art: Frank Miller /
Finished art: Klaus Janson

Bullseye: Daredevil's opposite, Elektra's killer.

Daredevil: The Man Without Fear limited series) Matt is holding back his inner rage and violence: "And the *guys*, they call me a *sissy*. It makes me *really mad*. It makes me want to *sock* them and I don't see any reason why I *shouldn't*." When Matt finally does hit one of them, "It was *great!*" Turning the other cheek is, in this story's moral scheme, considered to be a Bad Thing. Enraged that Matt hit someone else, the drunken Jack Murdock strikes Matt, and is immediately sorry. For Matt, however, the shock has turned his world upside down. "If even Dad can be wrong," the young Matt decided, "then anybody can do bad things. Anybody at all." Indeed, in Miller's *Daredevil*, even the hero himself is tempted to kill. As Daredevil tells a young boy caught up in a violent episode in an earlier Miller story, "We're only human . . . we can be weak. We can be evil. The only way to stop us from killing each other is to make rules, laws. And stick to them. They don't always work. But mostly, they do. And they're all we've got."

ELEKTRA SAGA

Daredevil #168 (1981) Script and pencils: Frank Miller /Inks: Klaus Janson

Daredevil #179 (1982) Script and art: Frank Miller / Finished art: Klaus Janson

Frank Miller named Elektra after Agamemnon's daughter in Greek mythology, who was obsessed with avenging her father's murder. Elektra Natchios met and fell in love with Matt Murdock when they were both students at New York City's Columbia University. Their romance came to an end when her father was accidentally killed; overwhelmed by grief she left for the Far East. In Miller's first issue as writer of Daredevil, she turned up in New York City as a modern-day female ninja operating as a ruthless bounty hunter. Soon the former lovers were on different sides: Elektra accepted the Kingpin's offer to become his new chief assassin and enforcer, replacing the villain Bullseye, who was in prison, and Daredevil was determined to bring her to justice for her killings. In a classic moment straight out of hard-boiled detective fiction, their old, intense feelings for each other resurfaced as they played out their roles as antagonists (above).

As the saga continued, *Daily Bugle* reporter and Daredevil confidant Ben Urich learned of the Kingpin's attempt to seize control of New York City through his pawn, mayoral candidate Randolph Cherryh. In a harrowing sequence set in a darkened movie theater, Elektra coldly threatened to kill Urich unless he dropped his investigation (above). Urich alerted Daredevil, leading to an explosive battle between Elektra and Daredevil (below), as much a nightmarish lover's quarrel as a battle over the city's fate, ending when she seemingly killed both Daredevil and the reporter.

Daredevil #179 (1982) Script and art: Frank Miller / Finished art: Klaus Janson

ELEKTRA SAGA (CONTINUED)

Left for dead, Daredevil and Urich descended into a modern-day underworld. Having learned that the Kingpin's wife, Vanessa, was still alive, a wraithlike figure wandering the sewers, the hero and his protégé went beneath the streets and found a civilization of savages led by a grotesquely massive brute calling himself the King, who had made Vanessa his queen (left). This king was a distorted mirror image of the Kingpin, ruling this underground parody of the city even as, above ground, on Election Day, the Kingpin made ready to take command of the real city. As if enacting a primeval rite, Daredevil overpowered the King, and Vanessa and the underground people pronounced him their new king. This victory against the Kingpin's primal double gave Daredevil the ability to thwart the real Kingpin in the world above. Daredevil blackmailed him into forcing Cherryh's withdrawal in exchange for Vanessa's return: again, Miller's Daredevil found he had to use immoral means to achieve a greater moral purpose.

Daredevil #181 (1982) Script and art: Frank Miller / Finished art: Klaus Janson

Daredevil #180 (1982) Script and art: Frank Miller / Finished art: Klaus Janson

Daredevil #181 (1982) Script and art: Frank Miller / Finished art: Klaus Janson

Meanwhile, Daredevil's most chilling nemesis, Bullseye, escaped from prison and sought to regain his position as the Kingpin's assassin. In a furious battle Elektra was mortally wounded by Bullseye (above right). Somehow she made her way to Murdock's apartment building and died in his arms (left).

Daredevil #181 (1982) Script and art: Frank Miller / Finished art: Klaus Janson

The vengeful Daredevil hunted down Bullseye and a deadly battle ensued. By the end Daredevil held Elektra's killer by one hand over a great height. Unrepentant to the last, Bullseye raised his weapon with his free hand; pushed to the edge of his morality, Daredevil let him drop. (Bullseye was not killed, but was left paralyzed for months.) But this was not enough to satisfy Murdock; daringly, Miller took him to operatic extremes of emotion, even digging up Elektra's coffin in his manic refusal to believe she was really dead.

Many issues later, Miller took another astounding leap: somehow, through sheer force of will and forgiveness, Daredevil purified Elektra's spirit. Her body, strangely preserved intact, disappeared, and the issue ended with the ethereal vision of Elektra, transcending her inner demons, reaching the top of a mountain that represented her quest for fulfillment. Was this what really happened to Elektra, or was it merely a metaphor for Matt coming to terms with his feelings for her? Miller did not say at the time, but the title of his later graphic novel *Elektra Lives Again* provided an answer: in this story line, Elektra and Bullseye slew one another, leaving Matt free of the two shadows that haunted and obsessed him. This story, however, has been relegated to the status of a "What If" story and is not part of the official continuity.

Daredevil #190 (1983) Script and pencils: Frank Miller / Inks: Klaus Janson

"And I guess that's what it all comes down to, Bullseye," Daredevil thinks in the final pages of issue #191. "When I hate you and your kind so fiercely I could cry . . . when it comes to that final, fatal act of ending you . . . my gun has no bullets." He was playing Russian roulette with an empty gun: "Guess we're stuck with each other, Bullseye."

MILLER'S RETURN

Miller surpassed all his previous *Daredevil* work when he briefly returned to the series in 1986 to write the seven-issue "Born Again" story line, handsomely illustrated by David Mazzucchelli, whose realism contrasts sharply with the neo-expressionistic artwork Miller had created for his *Daredevil* stories. Surprisingly, it is a work with explicitly religious themes, a tale of spiritual rebirth set in the Christmas season, with individual episodes bearing such titles as "Apocalypse," "Purgatory," and "Saved." Daringly, Miller utterly undercuts the genre tradition of the invincible super hero. The story chronicles the degradation, downfall, and ultimate redemption of Matt Murdock and his former girlfriend, Karen Page. Once again, the ostensible villain is the Kingpin, but the protagonists are really the victims of their own weakness, and are saved by the equally human impulses of compassion and forgiveness. In the end, Murdock publicly exposes the Kingpin as a criminal and establishes a new, simpler life with Page in the neighborhood of his childhood, Hell's Kitchen.

In super hero comics, one issue's victory over the villain provides merely a moment's rest before the next villain's attack in the following issue. Though Miller is accused of being the father of "grim and gritty" comics, "Born Again" nonetheless provides one of the few genuinely uplifting, triumphant, and well-earned "happy endings" in the super hero genre.

Miller's most recent return to the character was in the 1993 limited series *Daredevil: The Man Without Fear*, illustrated with cinematic energy by John Romita, Jr. This series stands as an anomaly in the Marvel canon, since it was the first time a major character's origin story was considerably revised. In the Lee-Everett origin Murdock took on the mask of power—his costume and role of Daredevil—in order to avenge a primal wrong, the slaying of his father. In Miller's new version, Matt wears a mask when he goes after the Fixer, but he dresses in his father's clothes, not the Daredevil costume. Accidentally killing a woman while pursuing the last of the Fixer's gang, Matt panics and runs away. Shaken by this awful accident, the young prankster in a ski mask was unable to come to terms with his "wild side" and thus unprepared for the role of Daredevil.

Enter Elektra, who introduces him to sexuality, which, in her case, is inextricably linked with danger and death. "This is where we *belong*," she tells him, "always on the *brink!* The *rest* lead safe, *numb* lives. . . . We're *two* of a *kind*. Drawn to the *edge* and *past* it." In contrast with previous views of her

BORN AGAIN

Daredevil #227 (1986) Script: Frank Miller / Art: David Mazzucchelli

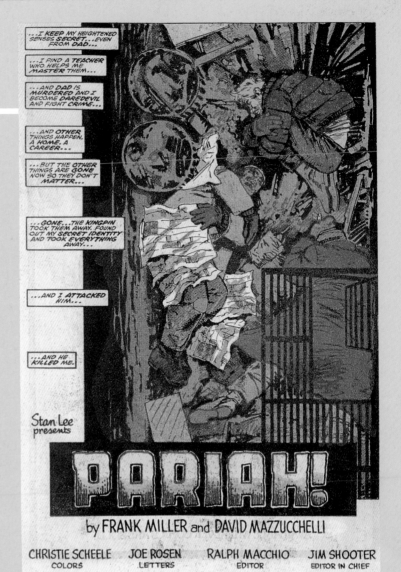

Daredevil #229 (1986) Script: Frank Miller / Art: David Mazzucchelli

"Born Again" began with Karen Page, Daredevil's former girlfriend, who had been written out of the series years before when she went to Los Angeles to pursue an acting career. Now a drug addict, Page desperately sold the secret of Daredevil's identity for enough money for a fix (above left); the slip of paper bearing his real name ended up in the hands of the Kingpin, who systematically proceeded to destroy Matt Murdock's life. Beset on all sides, Murdock fell into paranoia and madness; he finally confronted the Kingpin, who easily beat him and left him to die (above right). Wounded, he was nursed back to health in a homeless shelter. Meanwhile, Karen Page, torn by guilt, made her way back to New York, with a killer on her trail. In a brilliantly suspenseful sequence combining dynamic action and masterful crosscutting among many subplots, Miller and Mazzucchelli finally reunited Murdock and Page as he thwarted her would-be assassins and took back the mantle of Daredevil. Having conquered his inner demons, Matt showed Karen only forgiveness for her betrayal; he seemed to have risen above material things as well. As Daredevil, Murdock publicly exposed the Kingpin as a criminal and established a new, simpler life with Page in the neighborhood of his childhood, Hell's Kitchen (right).

Daredevil #233 (1986) Script: Frank Miller / Art: David Mazzucchelli

ELEKTRA: ASSASSIN

Elektra: Assassin # 4 (1986)
Art: Bill Sienkiewicz

In 1986 and 1987 Frank Miller teamed up with artist Bill Sienkiewicz to create *Elektra: Assassin*, an astonishing prequel to Elektra's first appearances in *Daredevil*. This limited series is pulp fiction transplanted into a postmodern world that mixes ancient mysticism with spies, cyborgs, and futuristic science. Intelligence agent John Garrett, slovenly, bigoted, and stupidly macho, was out to stop Elektra. Manipulated, thwarted, and humiliated by Elektra at every turn, and sexually attracted to her as well, Garrett gave up and the two joined forces against presidential candidate Ken Wind, the pawn of a demonic entity called the Beast that was out to cause a nuclear war that would devastate the globe. *Elektra: Assassin* was the first of Miller's over-the-top political satires, complete with Wind the demon-worshiper masquerading as a 1960s liberal and evil ninjas in the guise of yuppies. Were all of this depicted in a naturalistic style it would seem heavy-handed to say the least; what makes it wildly entertaining is Bill Sienkiewicz's extraordinary painted art, which ranges widely from photo-realism to outright caricature and is animated throughout by the artist's sharp sense of humor and palpable excitement at his own virtuosity.

Elektra: Assassin #5 (1986) Script: Frank Miller / Art: Bill Sienkiewicz

INSIDE THE LINCOLN MEMORIAL.

FOUR MONTHS AGO.

Elektra: Assassin #8 (1987) Script: Frank Miller / Art: Bill Sienkiewicz

Daredevil: The Man Without Fear #5 (1994)
Script: Frank Miller / Pencils: John Romita,
Jr. / Inks: Al Williamson
Matt Murdock as a grim vigilante on his
way to assuming the persona of Daredevil
in Miller's revised origin story.

Elektra: Root of Evil #2 (1995) Script:
D. G. Chichester / Pencils: Scott McDaniel /
Inks: Hector Collazo
Even though Miller killed her off in Elektra
Lives Again, Elektra lives on in the official
Marvel continuity and starred in the 1995
limited series Elektra: Root of Evil.

Daredevil #325 (1994) Script: D.G.
Chichester / Pencils: Scott McDaniel /
Inks: Hector Collazo
Artist Scott McDaniel also contributed
remarkable artwork to many of Daredevil's
recent adventures, for which he donned
this new armored costume, which he has
since abandoned.

college romance with Murdock, Elektra is now already clearly bordering on a madness "she can barely control"—that word again: she lures five would-be rapists into an alleyway and then brutally kills them just for her own amusement. After her father is killed, she leaves Matt, lest her potential for insanity consume him as well as her, and Matt feels "once again *punished*, for letting his *wild* side run free. For breaking the *rules*."

Murdock finally accepts his "wild side" and takes the name Daredevil when he directs his impulses into a more selfless mission than vengeance: that of rescuing a girl who has been abducted as a result of the Kingpin's machinations. Murdock wears a mask on this occasion, too, only donning the costume at the story's end, as he resolves to continue his new mission of fighting criminals who menace others and not simply those who have wronged him personally.

Later *Daredevil* writers have made their own significant contributions to the saga. Ann Nocenti and John Romita, Jr., created a very different femme fatale to ensnare the hero: the submissive, shy Mary Walker, who gave way to her other personality, that of the brutally violent and sexually daring Typhoid Mary. In his long run as *Daredevil* writer, Dan Chichester crafted "The Fall of the Kingpin," in which Daredevil finally evened the scales after "Born Again," playing a key role in toppling Fisk from his criminal empire. In later stories Chichester showed Fisk slowly and bloodily rising once more toward his former position of power, and infused elements of cyberpunk fiction into the series, pitting Daredevil against criminals manipulating the information superhighway. Recently, J. M. DeMatteis wrote the series, and began to repair Matt Murdock's fragmented psyche. To date, though, all subsequent *Daredevil* writers have been building upon Miller's reworking of the series; his mark on *Daredevil* has proved to be a permanent one.

Sgt. Fury and His Howling Commandos #50 (1967)
Pencils: Dick Ayers / Inks: John Severin

Stan Lee and Jack Kirby adapted their brand of action-adventure story-telling to war comics with *Sgt. Fury and His Howling Commandos.*

NICK FURY

By 1963 it was nearly two decades since World War II, yet there was still a considerable audience among the baby-boom generation for comics stories set in a war that ended before they were born. Seeking to infuse the war genre with the character-izations and sheer energy they had brought to super heroes, that year Stan Lee and Jack Kirby created *Sgt. Fury and His Howling Commandos,* which ran successfully under different creative teams well into the next decade.

At the center of the series was Sergeant Nick Fury, a character one suspects Lee and Kirby partic-ularly liked. He was the most real of all of the clas-sic Marvel heroes of the 1960s: a normal man without superhuman abilities, without anything glamorous in his background whatsoever. In fact Fury was a product of the Great Depression, raised on the Lower East Side of Manhattan by his wid-owed mother. Hardened by his upbringing, this gruff man nevertheless in his own way represented what Lee and Kirby saw as the spirit of America, as much as did that other wartime hero, Captain America. A natural rebel, Fury cared little for con-ventional military discipline, usually going unshaven and often unkempt, arguing with his commanding officer, the ironically nicknamed Captain "Happy Sam" Sawyer, flouting his commands whenever he thought it necessary to get his mission done. In the end, that was what was most important about Fury: he embodied what the series saw as the irresistible spirit of the ordinary American. Fury was unstoppable in combat and inexhaustible in his courage. However he twisted the rules the army set down for him, he never wavered in his patriotism.

Fury was leader of the Howling Commandos, a special unit of soldiers created by the Allied powers for missions behind enemy lines. (In fact, such an American unit should have been called rangers, but "commandos" sounded better; the "howling" referred to their trademark battle cry.) The series captured something of the feel of the Lee-Kirby super hero books in the dynamic action scenes and the larger-than-life victories that the Howlers contin-ually scored against the Axis powers, in sharp con-trast with the grim, more realistic battle tales of the DC war books of that period, such as *Sgt. Rock,* in which war was less a series of glorious victories than a world of misery and death.

In 1965, only two years after the Commandos' debut, Lee and Kirby moved Fury into a surprising new context. They had already introduced the Fury of the present in an issue of *The Fantastic Four:* he

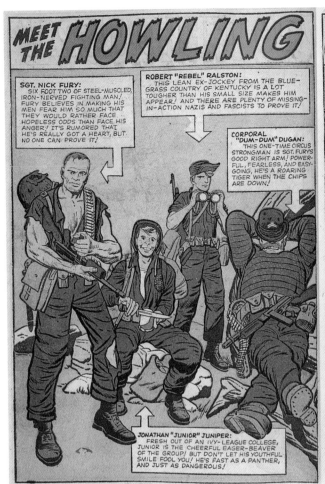

Sgt. Fury Annual #1 (1965)
Pencils: Dick Ayers / Inks: Frank Giacoia
As in war movies, the commandos consisted of colorful characters from diverse backgrounds, forming a microcosm of America: Irish-American (Dum Dum Dugan), African-American (Gabe Jones, Marvel's first black hero), southern white (Reb Ralston), New York Jewish (Izzy Cohen), Ivy League (Junior Juniper), and Italian-American (Dino Manelli, clearly based on Hollywood star Dean Martin), with an umbrella-toting Englishman (Percy Pinkerton, not seen here) to represent America's wartime partner.

was now a colonel and a special operative for the CIA. By *Strange Tales* #135 he also sported an eye-patch, a sign of an old war injury. Otherwise, despite his business suit, Fury was unchanged from the war. Now he was invited to become the director of S.H.I.E.L.D. (an acronym originally standing for the Supreme Headquarters, International Espionage, Law-Enforcement Division), an American intelligence and security force commanding an array of futuristic technology designed by Iron Man's alter ego, Tony Stark. The enemy S.H.I.E.L.D. opposed was a vast international subversive organization with its own high-tech arsenal, Hydra.

Obviously, Lee and Kirby were capitalizing upon the immense popularity of Ian Fleming's James Bond novels, the movies based on them, and the myriad imitations in the 1960s, like television's *The Man from UNCLE. Nick Fury, Agent of S.H.I.E.L.D.* replaced the sophisticated, debonair Bondian hero with a proudly uncultivated, earthy New Yorker: the unpretentious, plainspoken, often rude Fury. His

Sgt. Fury and His Howling Commandos #5 (1964)
Script: Stan Lee / Pencils: Jack Kirby / Inks: George Roussos (as "George Bell")
Although he made surprisingly few appearances over the series' run, Sgt. Fury's archvillain was Baron Wolfgang von Strucker, a man who was Fury's natural adversary: a Prussian aristocrat complete with monocle and dueling scar, who was already something of an anachronism in 1940s Germany, and who arrogantly regarded Fury as an upstart from the lower classes.

presence grounded Bond's world of superscientific fantasy in a credible human reality. As Fury himself realized, his situation really had not changed since the 1940s: once again, he was fighting a war, and he still charged into action himself, with torn shirt, gun in hand, and cigar clamped in his teeth, at every opportunity.

Unlike Bond's SPECTRE, Hydra was organized

Strange Tales Vol. 1 #156
(1967) Script and art: Jim
Steranko

Should Auld Acquaintance
Be Forgot: Baron Strucker
confronts Fury once again,
this time as master of the
neo-fascist army called
Hydra.

like a cult religion, its members garbed in green
robes, concealing their identities behind masks.
They worshiped their leader, the Supreme Hydra,
and continually repeated their mantra: "Hail Hydra!
Immortal Hydra! We shall never be destroyed! Cut
off one arm and two more will grow in its place!"—
a reference to the many-headed Hydra of Greek
mythology. Hydra was, in fact, Nazism recast in
stylized form by Lee and Kirby; their successor on
the series, Jim Steranko, clearly realized this when
he revealed the secret master of Hydra to be Baron
Wolfgang von Strucker himself, who had been
Fury's main nemesis in the earlier stories set during
World War II.

Steranko remained only a few years at Marvel,
but his spectacular work as writer and illustrator of
the *S.H.I.E.L.D.* series remains unsurpassed to this
day. His striking design sense, his mastery of cine-

matic staging and montage, his flair for vivid dialogue, and his genuine showmanship turned Fury's
activities into suspenseful, action-packed adventures
that equaled or surpassed the Bond movies that had
been their models.

As head of S.H.I.E.L.D. Fury has starred in several series over the years, but none has lasted for
very long. Perhaps the problem is that the
S.H.I.E.L.D. stories of the 1960s were so much an
outgrowth of the James Bond craze of that decade;
readers may no longer be as interested in such
flamboyant spy epics, and writers have not succeeded
in recapturing the flair and sheer energy that Lee,
Kirby, and Steranko gave the series. Also, as head of
a massive government agency Fury has by necessity
become an establishment figure, which may also
reduce his appeal to a young readership (most
Marvel super heroes operate on their own initiative).

BOND WITH AN EYE-PATCH

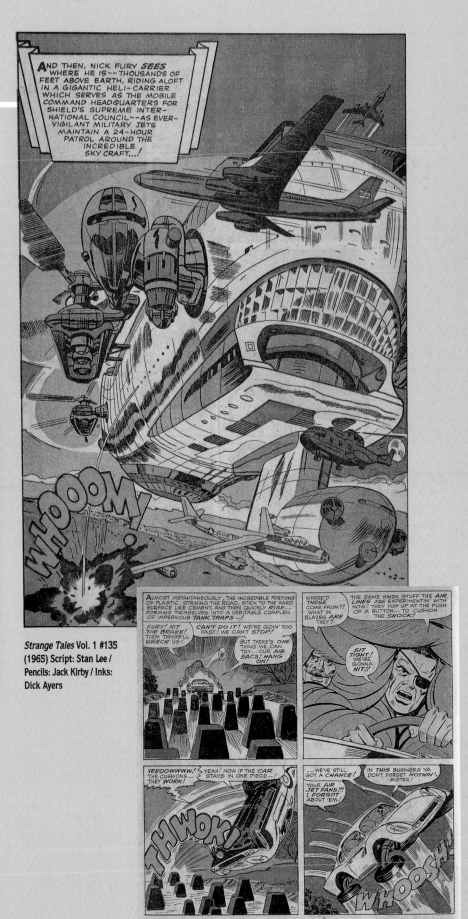

And then, Nick Fury *sees* where he is—thousands of feet above Earth, riding aloft in a gigantic heli-carrier which serves as the mobile command headquarters for SHIELD's supreme international council—as ever-vigilant military jets maintain a 24-hour patrol around the incredible sky craft...!

WHOOOM

Strange Tales Vol. 1 #135 (1965) Script: Stan Lee / Pencils: Jack Kirby / Inks: Dick Ayers

Strange Tales Vol. 1 #144 (1966) Script: Stan Lee / Designer: Jack Kirby / Pencils: Howard Purcell / Inks: Mike Esposito (as "Mickey Demeo")

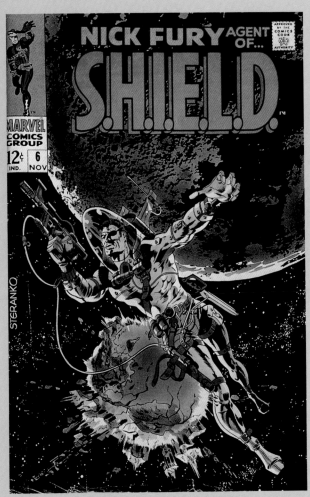

Nick Fury, Agent of S.H.I.E.L.D. Vol. 1 #6 (1968) Art: Jim Steranko

As head of S.H.I.E.L.D. Nick Fury found himself in a very different world than the battlefields of World War II. The massive heli-carrier that served as S.H.I.E.L.D.'s airborne equivalent of the Pentagon was nothing less than a space station miraculously aloft in Earth's gravitational field, and soon missions took the agency into outer space. Even Fury's S.H.I.E.L.D. flying car could perform tricks that Q never considered for James Bond's Aston-Martin: not only could it fly, but it came equipped with airbags.

Nevertheless, Fury has remained a popular character with Marvel's writers all these years and guest-starred in several series every month until mid-1995.

Fury has presented Marvel continuity with an interesting dilemma. In most cases comic book and strip characters do not age, or do so extremely slowly. Usually, the writers simply do not refer to past events that would date the characters: Little Orphan Annie, still a child today, does not make a point of talking about her exploits during the Great Depression of the 1930s. Similarly, the rule at Marvel is basically that, whatever year this is, it has been only roughly ten years since the events of *Fantastic Four* #1, and thus Spider-Man is still only in his twenties. Nick Fury, however, is rooted in World War II; years ago Walter Simonson and Howard Chaykin solved the dilemma of Fury's continuing vitality by explaining that after the war a scientist had subjected him to the "Infinity Formula," which has retarded his aging. So, ironically, thanks to his continuing visibility as a character for over thirty years, Marvel's leading non-superhuman ended up getting a superhuman ability after all: the secret of eternal youth. In 1995 Nick Fury's long career at Marvel seemingly came to an abrupt, unexpected end when the vigilante called the Punisher, who had been brainwashed by Fury's enemies, apparently shot him dead. Nick Fury, though, has made his life's work out of thwarting death, and, one way or another, we surely have not yet seen the last of this remarkable character.

Strange Tales Vol. 1 #165 (1968) Script and pencils: Jim Steranko / Inks: Frank Giacoia

Strange Tales Vol. 1 #167 (1968) Script and pencils: Jim Steranko / Inks: Joe Sinnott

JIM STERANKO

Jim Steranko's stint on S.H.I.E.L.D. reached its peak with two extraordinary thrillers. The first was Fury's one-man invasion of Hydra Island, which ended in finest James Bond style with his thrilling escape just as cataclysmic explosions sank it beneath the ocean. This was followed by Fury's clashes with the Yellow Claw, a villain in the Fu Manchu mold who had his own short-lived series in the 1950s and who turned out in Steranko's story to be one of Doctor Doom's robots. Here appeared two of Steranko's most exciting set pieces: an amazing aerial assault on the Yellow Claw's "Sky Dragon" headquarters by S.H.I.E.L.D. agents (left), and S.H.I.E.L.D.'s attack on the Yellow Claw in an astonishing four-page spread exploding with kinetic energy (below). The principal figures from left to right are Fury himself; the grinning S.H.I.E.L.D. agent Clay Quatermain (who seems to be Steranko's homage both to *King Solomon's Mines* hero Allan Quatermain and to movie star Burt Lancaster); Fury's love interest, the Contessa Valentina de la Fontaine (standing, firing a gun); S.H.I.E.L.D. agent Jimmy Woo, the hero of Marvel's 1950s Yellow Claw series (bending over a fallen body); Howling Commando turned S.H.I.E.L.D. agent Gabe Jones (being tackled); and the Yellow Claw himself (or, rather, Doctor Doom's robot in the Claw's image).

Marvel Premiere Vol. 1 #18 (1974) Pencils: Gil Kane / Inks: Dick Giordano
Marvel combined the martial-arts movies of the Far East with American super hero comics to create Iron Fist, a Westerner brought up in a latter-day Shangri-la and endowed with superhuman power.

PASSING TRENDS

Just as *Nick Fury, Agent of S.H.I.E.L.D.* tapped into the 1960s fascination with spy epics, so too Marvel over the years has often incorporated new trends in popular culture into its fictional mythos. The 1970s, for example, saw a growing interest in martial-arts adventure films, most importantly those of the late Bruce Lee. This led at Marvel to the introduction of two notable series. The first, oddly enough, resulted from the fact that Marvel had acquired the comics rights to Sax Rohmer's fictional villain Fu Manchu and his supporting characters. So it was that when Steve Englehart and Jim Starlin created the *Master of Kung Fu* series, they made their hero, Shang-Chi, the son of Fu Manchu. Shang-Chi's name means "the rising and advancing of a spirit," a phrase that aptly sums up the perennial theme of Englehart's work.

Master of Kung Fu achieved its full brilliance under the auspices of writer Doug Moench and a succession of talented artists, including Paul Gulacy, the late Gene Day, and Mike Zeck. Shang-Chi became the ally of Sir Denis Nayland Smith, working with him on missions for the British intelligence agency MI-6 and later for Smith's own independent outfit. Their exploits took them all over the world, combating a wide range of adversaries,

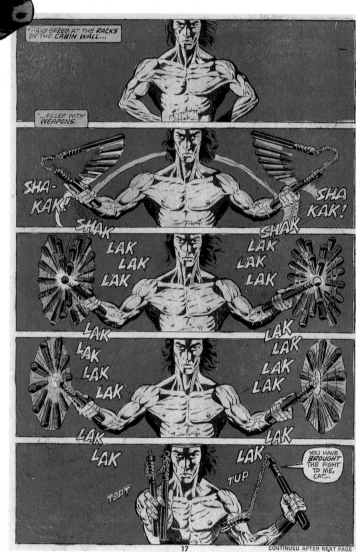

Master of Kung Fu #39 (1976)
Script: Doug Moench / Pencils: Paul Gulacy / Inks: Dan Adkins
Shang-Chi demonstrates his mastery of the martial arts.

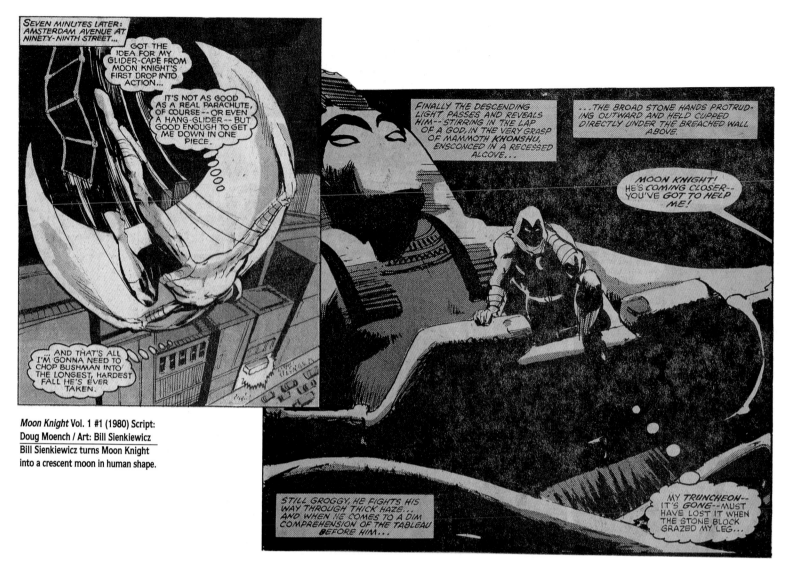

Moon Knight Vol. 1 #1 (1980) Script:
Doug Moench / Art: Bill Sienkiewicz
Bill Sienkiewicz turns Moon Knight
into a crescent moon in human shape.

Moon Knight Vol. 1 #28
(1983) Script: Doug Moench /
Art: Bill Sienkiewicz
Moon Knight awakes atop
a statue of Khonshu, the
ancient Egyptian deity who
is his patron.

ranging from realistic villains like drug dealers to exotic, even grotesque, warriors like Razorfist, who wore blades in place of his hands. The stories were somber and atmospheric, suited to the meditative calm of their hero, and contained elements reminiscent not only of Fleming but of John le Carré and even Milton Caniff's *Terry and the Pirates*. The high points would always be the inevitable reemergence of Fu Manchu himself. Moench portrayed Shang-Chi as a youth who began as an innocent, who sought to maintain his ideals, despite his disillusionments, in a corrupt world where both his allies and enemies played what he termed "games of deceit and death."

Marvel's other martial-arts hero fit more into the super hero mold. Created by Roy Thomas and Gil Kane, Iron Fist was Daniel Rand, a Caucasian American who was raised in the Himalayan land of K'un-L'un, a combination of Shangri-la and

Brigadoon, that appeared on Earth only once every ten years. In his late teens Daniel gained the power of the "iron fist" after slaying a dragon: by concentrating, he could strike with superhuman force. As Iron Fist, he returned to America to kill the man he held responsible for the deaths of his parents, only to renounce vengeance when he discovered the murderer had been reduced to a pitiable invalid. An innocent like Shang-Chi, Danny remained in America, reentering society and turning crimefighter.

His series is best remembered for the team that first collaborated on his later adventures: Chris Claremont and John Byrne. Not yet having achieved fame with the *X-Men*, these two were unable to save *Iron Fist* from cancellation, but he went on to appear for years as the unlikely partner of another uniquely 1970s character, Luke Cage, Power Man. Created in response to 1970s adventure films like

Punisher Vol. 2 #91 (1994) Script: Chuck Dixon / Art: Russ Heath Acclaimed war comics artist Russ Heath (whose past work has served as source material for Pop artist Roy Lichtenstein) turned his hand to the Punisher in this 1994 story line.

THE FEDS WERE WILLING TO LET HIM SLIDE ON CHARGES RANGING FROM MURDER TO STATUTORY RAPE.

ALL IN EXCHANGE FOR A LITTLE INFORMATION ON HIS BOSSES.

NONE OF WHICH WOULD PROBABLY HAVE LED TO ONE CONVICTION.

AND IF IT DID NOBODY WOULD SERVE ANY SERIOUS JAIL TIME.

Shaft, Cage was a streetwise African-American man who was unjustly imprisoned for murder. In jail he underwent an experiment that endowed him with superhuman strength and steel-hard skin. Escaping jail, he set himself up in an office in Times Square as a "hero for hire." Among his many memorable writers were Archie Goodwin and Don McGregor, and, after he gained a partner in *Power Man and Iron Fist*, Jo Duffy and Kurt Busiek.

Doug Moench's *Moon Knight* provided one of Marvel's stranger variations on the multiple-identity theme. Mercenary Marc Spector was left seemingly—perhaps actually—dead by his sadistic employer and then returned to life before the idol of an Egyptian moon god, Khonshu. Thus given a second chance, Spector decided to use his mercenary skills to make up for his amoral past. Soon thereafter he established himself in the New York area as a benevolent millionaire named Steven Grant while snooping for leads on criminals as the cab driver Jake Lockley and then fighting them in the guise of the costumed crimefighter Moon Knight. His ally, lover, and confidant was the beautiful Marlene Alraune: unlike more obsessive comic book vigilantes, Moon Knight had a healthy love life. Marlene aided him on numerous missions but worried that he was losing himself in his other identities. Ironically, she preferred him as Grant, which was not his real identity either, but then again, he was hardly the Spector of old; the mystery was exactly who Moon Knight was beneath all the role-playing.

Unfortunately, the concept of the series became overly convoluted under later writers, and Moon Knight was seemingly killed off in 1994.

Much of the fascination of the original *Moon Knight* series lies in following the evolution of artist Bill Sienkiewicz's work early in his career, as he makes the transition before our eyes from his early, sleek, glamorously realistic style influenced by Neal Adams to his first experiments with the expressionistic stylizations that have led to his successes with *The New Mutants* and *Elektra: Assassin*.

THE PUNISHER

The crimefighters mentioned above abided by the traditional code of super heroes: they cooperated with the law (for the most part), and they never killed. But one of Marvel's vigilantes, created in 1974, had a very different outlook. Relegated to supporting character status for decades, he finally became a star in the late 1980s and radically shifted the moral tone of mainstream adventure comics.

One day Vietnam veteran Frank Castle was having a picnic with his wife and two children in Central Park when they inadvertently witnessed a mob execution. (Why these gangsters chose to kill someone in broad daylight in Manhattan's foremost public space is beyond my ability to explain.) The assassins

then gunned down the Castle family, and Frank was the lone survivor. He identified the killers, but the authorities were unable to convict any of them. The deaths of his family unleashed in Castle a murderous, inextinguishable rage. He became the Punisher, an outlaw vigilante in a bulletproof costume bearing a death's-head design: Death himself, on a mission to hunt down and kill first those responsible for the slaying of his family, and then all other criminals he believed were beyond the reach of official justice.

The Punisher first appeared in 1974 as a supporting character in *The Amazing Spider-Man*, created by writer Gerry Conway and drawn by Ross Andru based on a visual design by John Romita, Sr.; his origin story was later filled out by Archie Goodwin. Over two decades the treatment of the Punisher has varied dramatically. No killer himself, Spider-Man could not approve of the Punisher's brand of vigilante justice, but writers Conway and Goodwin seemed to respect the Punisher's moral code, even if they disagreed with it. Frank Miller, on the other hand, did not allow Daredevil to tolerate the Punisher; the Punisher had overstepped the boundaries of morality and law, and Daredevil brought

SILVER SABLE AND DEATHLOK

For years Silver Sable, the glamorous European leader of the mercenary Wild Pack, created by Tom DeFalco and Ron Frenz, appeared coldly consumed by her profession of hunting down and killing criminals. Like the Punisher she was an emotional cripple, in her case compensating through her violent life for the pain caused her by the death of her mother and long disappearance of her father. But in the course of her own 1990s series, written by Gregory Wright, she came to realize how arid her life had become, and toyed with the idea of retirement before eventually returning to her leadership of the Wild Pack, whom she now regarded as her new surrogate family.

Deathlok Special #1 (1991) Script: Dwayne McDuffie and Gregory Wright / Pencils: Jackson Guice / Inks: Scott Williams

Silver Sable and the Wild Pack #1 (1992) Pencils: Steven Butler / Inks: Dan Panosian

Wright and fellow writer Dwayne McDuffie also explored the idea of the same person being both a nurturer and a warrior in the *Deathlok* revival of the 1990s. A forerunner of filmdom's *Robocop*, Marvel's original Deathlok, who debuted in the 1970s, was a soldier in a fictional future who had been converted into a cyborg, half-man and half-machine. Intriguingly, his human consciousness could mentally converse with the computer that now served as the other half of his brain. McDuffie and Wright's new version of Deathlok was a present-day pacifist and devoted husband and father whose human mind was transplanted into a robotic body designed for combat. Hence, the internal dialogues between the human and computerized sides of this Deathlok's brain became a metaphor for the division between a man's conscience and his amoral, destructive urges.

him to justice. Still later, writer Bill Mantlo had the Punisher go mad, even killing litterbugs, presumably as the end result of his belief in executing lawbreakers.

It was a landmark limited series by writer Steven Grant and artist Mike Zeck that set the tone for the Punisher's adventures ever since. Sane but ruthless, the Punisher garnered reader sympathy by taking on and defeating criminals far more vicious than himself, despite overwhelming odds that rendered him the underdog. From this point on the Punisher mounted in popularity; at his peak he starred in three monthly series: *The Punisher*, *Punisher War Journal*, and *Punisher War Zone*, in superbly crafted thrillers by such talents as writers Mike Baron, Chuck Dixon, Dan Abnett, and Andy Lanning and leading artists like Klaus Janson, Whilce Portacio, Jim Lee, and John Romita, Jr.

There is something nevertheless disturbing about the Punisher as a leading character. Stories clearly show the police and federal government out to capture him as an outlaw, while super heroes like Spider-Man and Captain America vow to bring him to justice. Moreover, the writers repeatedly demonstrate that Castle treats even his allies coldly and has no interests, or emotional attachments, or life beyond his endless war. Nonetheless, the very dramatic structure of the Punisher's stories casts him as a hero and thereby grants approval to his actions.

The Punisher's character is limited to acting out an obsessive compulsion to eliminate criminals violently. It is at once the source of his fascination and the cause of his spiritual damnation.

WESTERN HEROES

The Western genre rode high in comics from the 1940s into the early 1960s just as it did in movies and television, and Marvel had its own line of gunslinger heroes during this time. The Marvel archetype of the hero as alienated outsider was alive even then. Marvel's two leading Western stars, Kid Colt and the Rawhide Kid, were both outlaws, wrongfully wanted as murderers; like the mutant X-Men, they too were distrusted and hunted by the very citizenry they sought to protect. As the Western faded in popularity, Marvel's Western series increasingly looked like the more successful super hero books. The second version of the Two-Gun Kid even had a double identity: he worked as a lawyer but donned a mask to hunt down criminals, just like Daredevil. The Western heroes even got their own costumed criminals, like the Iron Mask, a nineteenth-century analogue to Doctor Doom. The Western hero who most resembled a super hero was the very first Ghost Rider, now known as the Phantom Rider, a masked vigilante who terrified his adversaries into thinking him a vengeful ghost, aided by luminescent dust that caused his costume and even his horse to glow eerily in the dark. Here, time-traveling Avengers encounter (from left to right) the Two-Gun Kid, Kid Colt, the Rawhide Kid, the Ringo Kid, and the Phantom Rider.

Avengers #142 (1975) Script: Steve Englehart / Pencils: George Perez / Inks: Vince Colletta

EIGHT

MUTATIS MUTANDIS: THE X-MEN

By the middle of the 1960s it was clear who the stars of the newborn Marvel Comics Group were: the list included Spider-Man at the top, with the Hulk and Fantastic Four and Captain America close behind. There was no way of predicting that by the late 1980s the popularity of all of the other 1960s characters would be far surpassed by that of a series that, although it too was a creation of the team of Lee and Kirby, had remained a second-string book throughout the 1960s. By the end of the decade the *X-Men* would be canceled as a commercial failure. But in the beginning of the 1990s the first issue of a new *X-Men* series sold a record eight million copies. The *X-Men* animated series dominated Saturday morning television ratings. If Marvel published only the books of the now extensive *X-Men* "family" of titles, it would still be one of the largest comics companies in North America.

The first issue of *X-Men* appeared in the same month in 1963 as *Avengers* #1; both were the work of Stan Lee and Jack Kirby; both were about teams of super heroes. But whereas the Avengers initially featured established solo stars, the X-Men was a team of teenagers that no reader had seen before. In fact, the X-Men could be described more specifically as a class of super heroes in training: if the Fantastic Four was a family, and the Avengers a club, then the X-Men was a school. They had a teacher who ran them through tests in the "Danger Room" and graded them, and they all wore the same basic costume, as if it were a school uniform.

This very concept seemingly restricted the X-Men to second-string status. Indeed, Kirby stopped penciling the series after issue #11, restricting himself to co-plotting and doing layouts for the stories. After another year he left altogether, and Lee departed the series soon afterward; they concentrated instead on the epic scale of the adventures in *Fantastic Four* and *Thor*.

Nevertheless, in their few years on the series, Lee and Kirby laid the conceptual groundwork for everything that followed. One key to the X-Men's potential was the same as Spider-Man's: the book was about the young, but the creators did not condescend to them. They were not called the X-Kids but the X-Men, and their doubts and frustrations were treated with respect and empathy. Although the X-Men's teacher, Professor Charles Xavier, a.k.a. Professor X, was a major presence in the series, the fact that he was disabled limited his role in combat: it was the teenagers themselves who had to fight and win their battles. The X-Men—five adolescents governed by a father figure—was a superheroic variation on a previously recurring theme in Kirby's work, the "kid gang," which he and Joe Simon had used in *Newsboy Legion, Boy Commandos*, and *Boys' Ranch* at other companies in the 1940s and 1950s.

A super hero's powers made him unique. There was only one Spider-Man, but this fact not only raised him above the rest of mankind; it also isolated him. He evoked the pain one suffers at feeling alienated from others and being misunderstood. With the X-Men Lee and Kirby took the next step: a person who feels alone with his or her "problem" longs for acceptance, for the company of others who share his or her outlook on life. The X-Men were a community of outsiders, who wanted to belong in the larger society but who, in the meantime, found comfort among each other. This is the second most important factor explaining the X-Men's audience appeal.

And the first? It was the fact that Lee and Kirby had come across the right dramatic metaphor for being "different" from other people: mutation. This was not a new idea. Mutation actually does take place in nature: it is one of the essential mechanisms of evolution. For a variety of reasons, a child may be born with a genetic trait not possessed by his or her parents or forebears. The child passes the trait on to his or her own offspring, and if the trait proves advantageous, it may continue to spread through the offspring's descendants.

In the real world most mutations are insignificant or harmful, but over the course of millions of years mutation gives rise to the evolution of new species. Throughout the 1950s and early 1960s there was growing public awareness of the dangers of atomic radiation, which was known to induce harmful mutations. The perils of mutation inevitably cropped up in the popular culture of the period, from the coming of the mutated dinosaur Godzilla to Superman's strange temporary mutations from exposure to Red Kryptonite.

In keeping with the times, early Marvel comics used radiation as the means of inducing super powers, as with the infamous "radioactive spider" whose bite transformed Peter Parker. Extending the theme

X-Men (First Series) #1 (1963) Script: Stan Lee / Pencils: Jack Kirby / Inks: Paul Reinman

X-Men (First Series) #1 (1963) Script: Stan Lee / Pencils: Jack Kirby / Inks: Paul Reinman

X-Men (Second Series) #1 (1991) Script: Chris Claremont / Pencils: Jim Lee / Inks: Scott Williams

The original X-Men—Cyclops, the Angel, the Beast, and Iceman—were teenage students at Professor Xavier's School for Gifted Youngsters, learning in secret how to use their mutant powers (and how to behave like adults). Their super hero costumes were actually combined school uniforms and workout clothes. They made their first appearance in issue #1 as Professor Charles Xavier telepathically summoned them to class (upper left). Later in the same issue, Jean Grey arrives and Xavier explains the name of his team to her (above). Like the Fantastic Four, the original X-Men were a surrogate family, with Xavier in the role of the outwardly austere but wise and caring father (left). Today the original X-Men are adults, the school is known as the Xavier Institute, and the entire team is more like mutant commandos honing their skills than mere students. Nonetheless, the familial bond between Xavier and his X-Men remains as firm as ever, and his bond with Jean (right) is especially strong.

X-Men (First Series) #3 (1964) Script: Stan Lee / Pencils: Jack Kirby / Inks: Paul Reinman

they had initiated with the Thing, Lee and Kirby showed that with the X-Men mutant powers were both a blessing and a curse: the mutations of three of the four male students in the original X-Men gave them special powers but also altered their physical appearance in strange ways. Bobby Drake, the Iceman, could freeze the air around him, sheathing his body in ice. Hank McCoy, the most intellectual of the group, had enormous hands and feet, which increased his agility to superhuman levels but gave him a somewhat apelike appearance that inspired his code name, the Beast. At least Hank still looked relatively normal in street clothes; Warren Worthington, the Angel, could pass as a normal human being only by strapping his huge, feathered wings into a concealed harness.

The unluckiest of the four was the leader of the five students, Scott Summers, who was unable to prevent force beams from shooting out from his eyes. The fact that the beams merged into a single ray, as if he had only one eye, gave Scott the code name Cyclops. Only the special ruby quartz lenses of the glasses he wore (and the visor he wore in costume) could hold back the beams. Emotionally repressed to begin with, Scott regarded himself as a potential danger to everyone around him, and especially to Jean Grey, the fifth student, the girl he secretly loved. Jean was the only one of the five who looked entirely normal when using her powers; as Marvel Girl she exercised telekinetic abilities that enabled her to move objects through willpower. In fact, she was portrayed as spectacularly beautiful; in the early 1960s, in keeping with the times, she was depicted as a goddess on a pedestal for whom the hero—Scott—longed. If he only knew that she secretly loved him as much as he did her!

Eventually it would be established that in the Marvel Universe most mutations did not manifest themselves until puberty. Hence mutation became a somewhat obvious symbol of the changes that overtake an individual moving from childhood into and through adolescence: both sexual awakening and the psychological adjustments to taking on greater independence and responsibility. The strange alterations in the bodies of the first X-Men, as well as many of the other mutants who would follow them, aroused the emotional and psychological unease and awkwardness that virtually all teenagers experience.

But it must be emphasized that mutation was also a blessing. Mutation not only made people look different, representing whatever the reader believed made him different from other people, but it also granted enviable, special abilities. Identifying with one of Marvel's super-powered mutants could let a reader feel that he might seem different from the people around him, but that that difference gave him an advantage and was something of which he could be proud.

MAGNETO

Even though a Marvel mutant rarely shared exactly the same powers with others, and in that sense was

X-Men (First Series) #111 (1978) Script: Chris Claremont / Pencils: John Byrne / Inks: Terry Austin

What Doctor Doom is to the Fantastic Four, Magneto is to the X-Men: the man who could have been their friend and ally, but who has instead become their foremost adversary.

unique, he or she was not one of a kind. A race of superhumans was evolving now, in the late twentieth century, which Lee and Kirby dubbed "Homo superior." It was like the religious idea of the Elect, a specially blessed segment of humanity destined for greater things than the rest. Here too was a fantasy of hope for society's outsiders: their numbers were growing, and they would become a force to be reckoned with.

Therein lay two major problems that have from the beginning provided the *X-Men* series with its dramatic impetus. First, how does this new breed of humanity utilize its growing strength? Charles Xavier taught his students to use their abilities to benefit humanity, even if the rest of humanity hated and distrusted them; Xavier was preaching the super hero genre's version of Christ's precept to turn the other cheek. Other mutants, however, disagreed.

A recurring theme in Marvel is the idea of the superrace that lives among us normal human beings and is divided into good and bad, like angels and devils warring over our fate. Lee and Kirby dealt with this idea later with the Inhumans and *Thor*'s New Men; Kirby would experiment further with the concept in his *New Gods* series for DC and *The Eternals* at Marvel. *The X-Men*, though, is where they first propounded this theme. The X-Men's role

in life was to protect humanity from the evil mutants, who sought not peaceful coexistence with "normal" humans but domination over them. And the X-Men's greatest enemy was the man later revealed to have once been Xavier's best friend, Erik Magnus Lehnsherr, better known as Magneto, who made his debut in the X-Men's very first issue.

Magneto interpreted the idea of "Homo superior" along the lines of the fascist concept of the master race. Commanding the force of magnetism, Magneto was himself nearly invincible, and he gathered around him a small terrorist band, the Brotherhood of Evil Mutants, as a sinister counterpart to the X-Men. Among them were his reluctant allies, the mutants Quicksilver and the Scarlet Witch, who would finally be revulsed by his tactics and join the Avengers; years later they would be revealed to be Magneto's son and daughter. Lee's Magneto continually thundered that humanity would take control of mutants unless the mutants struck first by taking over the governments of the world. Early issues of *X-Men* often centered on Xavier and Magneto's competing efforts to sway mutants they found to one or another side of the conflict.

The other problem created by the rise of the

X-Men (First Series) #14 (1965) Pencils: Jack Kirby / Inks: Wally Wood

THE SENTINELS

The Sentinels are a robotic secret police force, unquestioning, seemingly unstoppable, created to purge the mutant race from the world's population.

Marvels #2 (1994) Script: Kurt Busiek / Art: Alex Ross
As Kurt Busiek and Alex Ross's *Marvels* points out, the general public could accept people like the Fantastic Four, normal human beings who had gained superhuman abilities through accident, as heroes, because they were "one of us." But the mysterious process by which an incomprehensible fate created mutants in their midst, people who were "different" from the time of their births, spread fear and unease among the public.

Marvels #2 (1994) Script: Kurt Busiek / Art: Alex Ross

mutant superrace was the reaction of the mass of humanity to the superbeings in their midst. The earliest Lee and Kirby *X-Men* issues primarily concern the internecine wars between the good and evil mutants. But soon there was a sequence in which one of Magneto's underlings, the disguised Toad, demonstrates his superhuman athletic abilities in public, triggering a riot among the spectators from which the X-Men have to rescue him. This, however, was an isolated incident. Lee and Kirby finally fully addressed the issue of man's persecution of mutants with the introduction of the Sentinels. In this story line (issues #14–16) Professor Xavier agrees to a television debate with Dr. Bolivar Trask, who has been spreading fear that the unknown mutants among us will soon rise and reduce "normal" humanity to slavery. During the broadcast Trask unveils his line of defense against a mutant takeover: gigantic Sentinel robots, programmed to seek out, capture, and destroy mutants—a science-fiction version of an all-powerful secret police force dedicated to wiping out a minority group, in this case, mutants. Indeed, the Sentinels prove too powerful for Trask to control, and leave the studio, determined to save humanity from the mutants even if humanity itself must be conquered in the process.

In retrospect, the Sentinels story was a turning point in the Marvel Universe, adding layers of complexity to what had been largely a genre in which super heroes beloved of humanity fought and vanquished super-villains hated by all. What had been one motif in the adventures of super heroes like Sub-Mariner and the Incredible Hulk—that a "hero" could also seem to be a menace—suddenly became a major theme. In *Marvels*, Busiek dramatizes the change by setting the public revelation of the Sentinels on the very same day as Reed Richards and Sue Storm's wedding. The celebration of superhumanity in the daylight gives way to a terrifying witch-hunt after dark.

The X-Men stopped the Sentinels on this occasion, but this robot militia and the undying prejudice they embody have remained key elements of *The X-Men* mythos ever since (the Sentinels were the menace in the first episode of the *X-Men* animated series). If an individual mutant could represent any individual person who felt alienated from society.

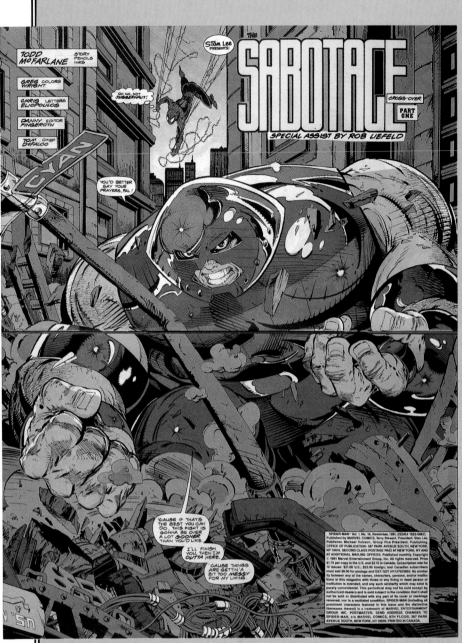

Spider-Man #16 (1991) Script and art: Todd McFarlane

JUGGERNAUT

Returning once more to their theme of rival siblings, Lee and Kirby gave Xavier a resentful half brother, the biblically named Cain Marko, who stumbled across a mystical gem in an Asian temple and became transformed into the unstoppable Juggernaut. Ultimately, however, the Juggernaut's brute strength and even more brutish rage proved ineffectual against the force of Xavier's mind and his strength of character and determination: having smashed through the X-Men's physical defenses, the Juggernaut collapsed like a rag doll before the seemingly helpless man in the wheelchair. Of course, like other popular Marvel villains, he has returned many times to do battle with the X-Men. This illustration by Todd McFarlane, better known for his Spider-Man comics, comes from a crossover series.

No mere words of ours can do justice to the fury of Ka-Zar's attack ... so we'll attempt no such written description!

X-Men (First Series) #10 (1965) Script: Stan Lee / Pencils: Jack Kirby / Inks: Chic Stone

X-MEN, WELCOME TO--

--THE SAVAGE LAND!

Uncanny X-Men #114 (1978) Script: Chris Claremont / Pencils: John Byrne / Inks: Terry Austin

MARVEL COMICS GROUP PRESENTS: AUG 75¢ u.k. 35p #17

KA-ZAR THE DETECTIVE

TAG, YOU'RE IT!

WITH KEVIN PLUNDER · SHANNA O'HARA AND INTRODUCING ZABU AS "REX"

Ka-Zar the Savage #17 (1982) Art: Brent Anderson

Lee and Kirby sent the X-Men to the Savage Land, a pre-historic jungle complete with dinosaurs hidden somewhere on the Antarctic Continent. (A ring of volcanoes and alien climate-control devices maintain the Savage Land's tropical temperatures.) Here they encountered the jungle's ruler, Ka-Zar (center), eventually to be revealed as a British nobleman who had been marooned in the Savage Land as a child. His name taken from a Timely character of the 1940s, Ka-Zar was clearly inspired by Edgar Rice Burroughs's Tarzan, and his jungle is a modern-day prehistoric realm in the fictional line leading from Sir Arthur Conan Doyle's *The Lost World* to Michael Crichton's *Jurassic Park* (left). It is inhabited by creatures from the age of dinosaurs to the Ice Age, including Ka-Zar's saber-toothed tiger Zabu. Originally speaking like the monosyllabic Tarzan of the movies, in later appearances in various Marvel series Ka-Zar took on the more sophisticated manner of the aristocratic Tarzan of the novels. But it was in Ka-Zar's own series in the 1980s that writer Bruce Jones and artist Brent Anderson broke away from the Burroughs noble savage mode. In their ironic take on the Tarzan myth they cast Ka-Zar and a newer jungle character, Shanna the She-Devil, as a very contemporary couple (right) who chose to return to life away from civilization but were still subject to the ups and downs of modern relations between the sexes.

After the *X-Men* series was canceled in 1970, the team members popped up from time to time over the next half-decade in other Marvel series, but to no lasting effect. The Beast won his own short-lived series, in *Amazing Adventures*, in which he rather foolishly drank a mutagenic serum of his own creation and ended up a creature far more worthy of his name, with blue fur, pointed ears, and fangs. Soon, Steve Englehart wrote him into the Avengers, where he became a witty, fun-loving teddy bear turned super hero. Although he was transformed back to his original, more human appearance for a time, it is the blue, furry Beast who has remained most popular with Marvel readers and writers.

then mutants in general could serve as a metaphor for any segment of humanity that was the victim of prejudice and oppression for reasons of race, religion, gender, or any other cause. Indeed, explicit analogues to the fates of various real-life oppressed minorities have been made over the years in *The X-Men*, resulting in some of their strongest, most memorable story lines, as shall be seen.

THE YEARS OF TRANSITION

Lee and Kirby's successors on *The X-Men* were Roy Thomas, the first major writer brought into Marvel from the ranks of the company's growing fandom, and romance comics artist Werner Roth. Although not a master of super hero action scenes, Roth brought a generally underrated sensitivity to the evolving romance between Scott and Jean, as well as to other subplots. Roy Thomas clearly relished the opportunity to write a super hero team book of his own and quickly proved himself a worthy successor to Lee. Still, the book's sales faded, and drastic measures were called for. Thomas killed off Xavier and sent the X-Men off as individuals or pairs for their own adventures as independent adults; this tactic did not work, the team was reunited, and it would eventually be demonstrated that Xavier was still alive and that it was a shape-shifting double who had perished.

Perhaps the original *X-Men*'s greatest burst of glory came at the very end of its run, with a series of collaborations between Thomas and artist Neal Adams, whose idealized figures, spectacular settings, and cinematic flair for storytelling made him the most influential comic book artist of the late 1960s and early 1970s. Finally, *The X-Men* acted like a flagship title, sending its heroes all over the world, battling the immense Living Monolith in Egypt, having a showdown with the invincible Sentinels, tracking down Magneto in his new base in Ka-Zar's Savage Land, clashing with the mutant Sunfire in Japan, and driving back the invasion of the alien Z'nox. These were astonishingly impressive adventure tales, but they came too late to save the series, which ended in 1970 with issue #66.

SECOND GENESIS

Although Roy Thomas and other fans turned Marvel writers were fond of the X-Men, it seemed impossible to revive them successfully on an ongoing basis. So Thomas decided to revamp the concept: there would be a new team of mutants, drawing its members from different countries. Writer Len Wein and

NEAL ADAMS

X-Men (First Series) #58 (1969) Script: Roy Thomas / Pencils: Neal Adams / Inks: Tom Palmer

Neal Adams collaborated with writer Roy Thomas on several episodes of the Kree-Skrull War in *The Avengers* and a lengthy run of X-Men adventures, endowing them all with dynamic, fast-paced action and spectacles that were both believable and far larger than life. In the unusual double-page spread above, Adams uses tilted perspectives, slashing, diagonal panel borders, and panels resembling wide-screen movies to heighten the fury of a battle between Iceman and a Sentinel, while a television interview serves as counterpoint, voicing the racial hatred that led to the Sentinels' creation. Right, Adams ably conveys a sense of the sheer power in Cyclops and his optic blast.

X-Men (First Series) #59 (1969) Script: Roy Thomas / Pencils: Neal Adams / Inks: Tom Palmer

artist Dave Cockrum took this inspiration and brought it to fruition in *Giant-Size X-Men* #1, published in 1975. In this tale the original X-Men were captured by Krakoa, a jungle island that proved to be a colossal sentient mutant plant colony, and Professor Xavier hurriedly recruited a new team of mutants to rescue them. Among them were two former adversaries of the X-Men, Sunfire, who quit immediately after their first mission, and Banshee, a middle-aged Irishman who could create intense sonic force by screaming. Other newcomers included Peter Rasputin, alias Colossus, a young Russian farmboy in his late teens who could transform into a superhumanly strong being with "organic metal" skin; Kurt Wagner, known as Nightcrawler, a German circus performer resembling a demon, complete with a tail, who could teleport himself, seeming to disappear in a puff of smoke; Ororo Munroe, code-named Storm, an African-American born with blue eyes and white hair, whose parents were killed in Egypt when she was a child, leaving her to grow up on the streets of Cairo as a pickpocket, and who eventually made her way to Tanzania, where her mutant ability to control the weather caused her to be worshiped as a goddess; John Proudstar, alias Thunderbird, a superhumanly strong Native American who, to the surprise of the readers, was killed off on his second mission; and Wolverine,

otherwise known only as Logan, a short, feisty Canadian with retractable claws who had defected from his job with Canadian intelligence to join Xavier's team.

The rescue mission proved successful, so much so that most of the original X-Men felt little guilt about quitting the team to go out into the world and find independent lives for themselves. The with-drawn Cyclops stayed behind as field leader for Xavier's new team, and the *X-Men* series was relaunched in August 1975.

Giant-Size X-Men #1 (1975)
Script: Len Wein / Art: Dave Cockrum
Professor X reviews the original lineup of the "new" X-Men. Facing him, from left to right, are Colossus, Storm, Thunderbird, and Nightcrawler. Standing on the staircase, from left to right, are Banshee, Sunfire, who can project intense heat, and Wolverine in his original costume. Cyclops served as the team's leader in the field.

X-Men (First Series) #101 (1976) Art: Dave Cockrum

By issue #100 Claremont and Cockrum were capping off a war aboard a space station between the new X-Men and the re-created Sentinels with the heroes' desperate escape back to Earth through a solar radiation storm. Jean Grey, who had been captured by the Sentinels, knowingly faced death by serving as pilot of the escape craft, since the cockpit would be exposed to the lethal rays. Issue #100 ended with her seeming death in the radiation storm; in issue #101 the escape ship crashed into the waters off New York, and Jean—or what seemed to be Jean—rose from the waters, reborn with greater powers. She named herself Phoenix, after her resurrection and the energy that manifested itself around her body in the shape of a fiery bird of prey.

insightful skill in characterization, a flair for dramatic dialogue, and a genuine knowledge and feel for science fiction. He revolutionized the treatment of women in super hero comics: too often in the comics of the 1960s the heroines seemed reduced to standing on the sidelines and pointing in combat (like the Scarlet Witch) or hiding (like the Invisible Girl). Claremont, however, was skilled in portraying independent-minded, strong, passionate female characters who were truly equal partners with the men in the books and influenced the female characterizations throughout the genre. Moreover, the *X-Men*'s women characters in both the comics and now the animated series won it an unusual degree of popularity among female readers and viewers, who usually evince little interest in the male-dominated super hero genre.

Not since the Lee and Kirby *Fantastic Four* had a single series been as prolific as the Claremont *X-Men* in creating new characters. Over his decade and a half on the X-books, Claremont and his collaborators added popular new heroes to the original team and devised an array of compelling villains, including the Shadow King, Xavier's first mutant opponent, who existed as pure consciousness that took over living host bodies, and the members of the Hellfire Club, an actual eighteenth-century organization updated into a cabal of wealthy, decadent, mutant businessmen secretly manipulating the government and the world economy, as well as two more teams of mutant heroes: the New Mutants and Excalibur.

Busy with other writing assignments, the "new" X-Men's co-creator Len Wein soon left the series: he plotted the first two regular issues but left the scripting of the dialogue to a newcomer, Chris Claremont, who worked at first with artist Dave Cockrum. Claremont would stay for sixteen years, writing every issue of the regular *X-Men* series, as well as limited series, special annual issues, and three spin-off series. This is a record no other Marvel writer has matched.

Claremont's strengths as a writer were many:

Uncanny X-Men #129 (1981) Script: Chris Claremont / Pencils: John Byrne / Inks: Terry Austin

Claremont and Cockrum were sparking great interest among discerning comics readers. Only six issues into their run, they devised a memorable story line involving Jean Grey and the re-created Sentinels, which ended with her seeming death and rebirth as the powerful Phoenix. Over the following year they transported the X-Men to the empire of the Shi'ar, their answer to Lee and Kirby's Kree and Skrulls, whose mad ruler was about to unleash the power of the M'Krann Crystal, which contained forces sufficient to obliterate the cosmos. In a dazzling finale Phoenix entered the crystal and, linking herself with the spirits of her fellow X-Men, used her new command of cosmic forces to knit the unraveling fabric of reality back together. It was an ending that evoked a sense of wonder and amazement unseen before in *The X-Men*.

Actually, Cockrum had left the series only an issue before this denouement. His successor was a new Canadian comics artist, John Byrne. He had already collaborated with Claremont on the martial-arts series *Iron Fist*, and now he began co-plotting *The X-Men* with him. With Byrne and Claremont teamed on the book, the *X-Men* entered a new period of greatness.

Although Byrne and Claremont often differed sharply on characterization and story lines, they succeeded for three memorable years of *The X-Men* (issues #108–143) in blending their respective talents together to create a run of issues that rivaled Lee and Kirby at their best. Claremont and Byrne made their considerable reputations with comic-

book readers with this series, as did their admirable regular collaborators, inker Terry Austin, colorist Glynis Oliver, and letterer Tom Orzechowski.

The high point of this period was the 1980s "Dark Phoenix" saga in what was now titled *The Uncanny X-Men* (issues #132–137), an epic that successfully ranged from intimate characterization to the destruction of a planet. Only a few months after this triumph, Claremont and Byrne produced another classic that may have had even greater impact on the Marvel Universe. "Days of Future Past" (issues #141–142), which established both themes and storytelling techniques that predominate in the *X-Men* to this day, presents a possible twenty-first-century future in which the federal government has unleashed the Sentinels against the increasing mutant population with disastrous results.

Uncanny X-Men #142 (1981) Pencils: John Byrne / Inks: Terry Austin

Byrne introduced the teenager Katherine "Kitty" Pryde, who discovered at age thirteen that she could walk through walls (above left, with Storm). Much younger and more innocent than the rest of the now-adult team, Kitty provided a link to the X-Men's past as school kids, her high spirits compensating for the darkness that had suffused the series. Under Claremont's handling she evolved and matured, developing a youthful romance with the older Colossus that ended painfully, discovering a talent for working with computers (one of the numerous ways Claremont anticipated and incorporated coming trends in his stories), and forming an unexpected bond with an unlikely mentor, Wolverine.

DARK PHOENIX

Throughout his career Claremont has been fascinated with the power of the repressed shadow side of the human personality. Having presented Phoenix as the ultimate expression of the life-giving side of Jean's personality, he now turned his attention to the Phoenix as a force of destruction. Opening with the long-postponed consummation of the romance between Scott Summers and Jean Grey (or, rather, the Phoenix, bearing part of Jean's consciousness, below), the story quickly put an end to the lovers' newfound bliss. Mastermind, a former member of Magneto's Brotherhood, had sought to gain induction into the Hellfire Club's ruling Inner Circle by utilizing his powers of illusion to gain control of Jean's mind. When the X-Men invaded the club's New York headquarters, Mastermind sprung his carefully constructed trap, making Jean hold the other X-Men captive, including her lover Cyclops (opposite top left). Fighting free of Mastermind's control, she wreaked a terrible vengeance on him. Released from all inhibitions and repossessed of her own free will, the insane Dark Phoenix proved unstoppable, leaving Earth and destroying both an inhabited planet (opposite top right) and a Shi'ar starship. She finally engaged in a battle of wills with Xavier, who telepathically managed to reach Jean's buried normal personality and join with it to exorcise Dark Phoenix's madness.

But that was not the end. The enraged Shi'ar arrived and demanded that the now "normal" Jean be executed lest she menace the cosmos once again. Xavier demanded a trial by combat to determine Jean's fate, and the X-Men were teleported aboard a Shi'ar starship, commanded by the Empress Lilandra (opposite center, from left to right: Cyclops, Phoenix, the Beast, Angel, Nightcrawler, Wolverine, Storm, Colossus, Professor X, the Shi'ar's foremost warrior Gladiator, Empress Lilandra, and her prime minister Araki). The battle was then fought on the Blue Area of the moon (opposite below left) between the X-Men and the Shi'ar Imperial Guard, with Scott and Jean fighting side by side, united by their love for one another. Originally Claremont had intended that Jean would live, but that the Shi'ar would strip her of her powers, and that she and Scott would then leave the team to lead a new life together—it would be a bittersweet ending to the story of their love. Editor-in-chief Jim Shooter, however, believed that excusing Jean for Dark Phoenix's actions on the basis of temporary insanity was insufficient: Dark Phoenix's crimes required a full expiation. And so, at the high point of the battle, Jean found herself reverting to Dark Phoenix once again. Unable to prevent the tide of madness from washing over her, Jean committed suicide before Scott's horrified eyes (opposite below right). With this new, stunning ending, this tale of self-sacrifice surpassed the limits of super hero melodrama, taking on truly tragic dimensions.

Uncanny X-Men #136 (1980) Script: Chris Claremont / Pencils: John Byrne / Inks: Terry Austin

Uncanny X-Men #132 (1980) Script: Chris Claremont / Pencils: John Byrne / Inks: Terry Austin

Uncanny X-Men #132 (1980) Script: Chris Claremont / Pencils: John Byrne / Inks: Terry Austin

Uncanny X-Men #135 (1980) Script: Chris Claremont / Pencils: John Byrne / Inks: Terry Austin

Uncanny X-Men #137 (1980)
Script: Chris Claremont / Pencils:
John Byrne / Inks: Terry Austin

"DAYS OF FUTURE PAST"

Uncanny X-Men #141 (1981) Pencils: John Byrne / Inks: Terry Austin

In "Days of Future Past" Claremont and Byrne envisioned a future where Xavier's dream turned to nightmare. The assassination of Senator Robert Kelly, a leader of the antimutant movement, by the terrorists of a new Brotherhood of Evil Mutants, led to the U.S. government unleashing new Sentinels. Deciding it was necessary to protect humankind, the Sentinels conquered the United States. After years of warfare, Xavier was killed, his mansion destroyed, and those mutants who survived were incarcerated in concentration camps. (In one dramatic panel, below left, Sentinels, having just slain Franklin Richards, the mutant son of Reed and Sue, pursued other X-Men fleeing captivity in this alternate future New York City.) In the New York area camp Magneto, now, ironically, confined to a wheel-chair, led what remained of the X-Men. There, Rachel Summers, the daughter of Scott Summers and Jean Grey in this alternate reality, used her mental powers to send the spirit of another X-Man, Kate Pryde, back through time into the body she had decades before, when she was Kitty Pryde, a new thirteen-year-old recruit to the X-Men. Kate's assignment was to get the X-Men to prevent Kelly's assassination and thus create a new and better future. The story grippingly crosscut between the X-Men's efforts to thwart the Brotherhood in the present and their attempt to escape from the Sentinels' prison camp in the future. One by one the remaining future X-Men were slain before the readers' eyes—Wolverine's annihilation, below, leaving only his indestructible adamantium-infused skeleton—and only Rachel, tending the elder Kate's body, remained to discover whether the X-Men in the past will succeed in altering the future. The assassination was prevented, but there was no easy catharsis: whether or not the X-Men's world of the present would still evolve into Rachel's nightmarish future remained unresolved.

Uncanny X-Men #141 (1981) Script: Chris Claremont / Pencils: John Byrne / Inks: Terry Austin

Uncanny X-Men #142 (1981) Script: Chris Claremont / Pencils: John Byrne / Inks: Terry Austin

224

"QUESTION IS: CAN I KILL WOLVERINE BEFORE HE CAN REACH ME AN' CUT ME INTO SHISH-KEBAB WITH THOSE FREAKY CLAWS OF HIS?"

WELL, BUB, WOLVERINE IS VIRTUALLY UNKILLABLE.

WOLVERINE'S CLAWS ARE ADAMANTIUM, THE STRONGEST METAL KNOWN -- CAPABLE OF SLICIN' THROUGH VANADIUM STEEL LIKE A HOT KNIFE THROUGH BUTTER.

AN' FIVE METERS O' FLOOR AIN'T MUCH DISTANCE AT ALL -- FER ME.

Uncanny X-Men #133 (1980)
Script: Chris Claremont /
Pencils: John Byrne / Inks:
Terry Austin

Wolverine was an example of a new, harder-edged kind of hero emerging in 1970s popular culture. Here Claremont links Wolverine with Clint Eastwood's movie persona through an homage to *Dirty Harry.*

WOLVERINE

It was Wolverine who proved to be the "new" X-Man who most strongly struck a chord with readers. He was originally created by Len Wein and artist Herb Trimpe for an issue of *The Incredible Hulk*, following a suggestion by Roy Thomas and a sketch by John Romita, Sr. Wolverine was shorter than other heroes and the seeming underdog when matched against larger opponents. His feistiness more than made up for his height, however, and perhaps if Wolverine had been bigger, he would have seemed a bully. After all, he was bad tempered and had a new kind of weapon—his razor-sharp claws. This represented a major escalation in the level of acceptable violence back in the days when super heroes did not use bladed weapons, much less guns.

Wolverine's claws and skeleton were reinforced with the Marvel Universe's fictional, nearly indestructible metal adamantium, and Claremont would eventually reveal that Wolverine's principal mutant power was his ability to recover with superhuman rapidity from lethal injuries. This power was the equivalent of a super hero's traditional invulnerability for a time with more violent tastes in popular culture. Bullets would bounce off Superman's chest harmlessly; bullets would wound Wolverine, cause him pain, and make him bleed, but within minutes he would be fully recovered.

In his first appearances during Dave Cockrum's original run on *The X-Men*, Wolverine was often made to look somewhat foolish, charging into battle only to get the worst of it. But Claremont was already establishing that Wolverine was far more than a tough-talking blowhard: in fact, he had to struggle constantly to prevent himself from slipping into berserker rages, a tendency he had fought for years, seemingly to no avail.

In part because they both hailed from Canada, John Byrne took a particular interest in the character, and together he and Claremont treated Wolverine far more seriously, portraying him as a self-assured master at fighting, never shrinking from whatever violence was necessary to get the job done, somewhat reminiscent of Clint Eastwood's cinematic heroes. Wolverine was, as his trademark catchphrase went, "the best there is at what he does."

In effect Wolverine served the same role in the "new" X-Men that the Thing had in the Fantastic Four and Hawkeye had in the Avengers in the 1960s. All three were angry rebels, challenging the team's authority figures, disruptive of the team's harmony, ids that were ultimately kept in place by the superego figures, whether Reed Richards, Captain America, or Professor X and Cyclops. But the ante was being raised with each decade: the Thing and Hawkeye were hotheads, but only Wolverine was a potential killer.

A turning point came in a story by Claremont and Byrne set in Ka-Zar's Savage Land, in which Wolverine seemingly slew one of the villain's guards

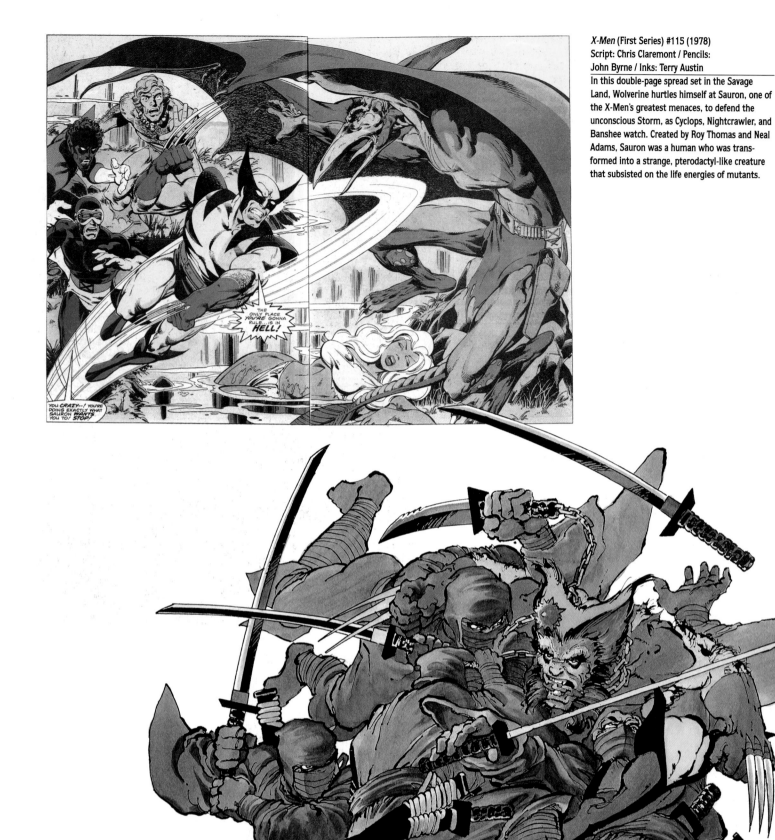

X-Men (First Series) #115 (1978)
Script: Chris Claremont / Pencils:
John Byrne / Inks: Terry Austin
In this double-page spread set in the Savage
Land, Wolverine hurtles himself at Sauron, one of
the X-Men's greatest menaces, to defend the
unconscious Storm, as Cyclops, Nightcrawler, and
Banshee watch. Created by Roy Thomas and Neal
Adams, Sauron was a human who was trans-
formed into a strange, pterodactyl-like creature
that subsisted on the life energies of mutants.

Art: Frank Miller

Chris Claremont and Frank Miller's
Wolverine limited series clearly
defined Wolverine's personality for all
subsequent stories: the man striving
for rationality and humanity through
his struggle to control and restrain
the beast within himself.

Marvel Comics Presents #84 (1992)
Script and art: Barry Windsor-Smith
Best known for his elegant work on
the early issues of Marvel's *Conan the
Barbarian*, artist Barry Windsor-Smith
here instead powerfully captures the
animalistic savagery within Wolverine.
This excerpt from the "Weapon X"
story line shows Logan soon after the
experiment that infused his skeleton
with adamantium and drove him insane.

off-panel. Today's readers would take such a scene for granted, but in the late 1970s super heroes never killed. For better or worse, a line had been crossed and something had to be done.

The answer lay in Japan. In a story set there, Claremont and Byrne introduced Wolverine's fascination with Japanese culture, which offered the self-discipline and code of honor to which Wolverine aspired. They also introduced Mariko Yashida, a young Japanese woman who embodied the beauty and serenity missing from Wolverine's life and with whom he immediately fell in love. It was another reworking of the Beauty and the Beast theme, like the romance between the Thing and Alicia, although this time the "beast" suffered not from physical grotesquerie but from raging psychoses.

Wolverine's hair-trigger capacity for rage lasted as long as Byrne remained on the *X-Men*; after his departure—like Kirby before him, he now preferred to write his own stories and after completing issue #143 he moved on to *The Fantastic Four*—Claremont took the final major step in developing Wolverine's personality into what it is today. In the *Wolverine* limited series, drawn by Frank Miller, Wolverine returned to Japan, where he clashed with Mariko's physically abusive father, the crimelord Shingen. In the grip of his berserker fury, the animalistic Wolverine was humiliatingly bested by the old man in personal combat. Not until Wolverine mastered the animal within himself through rational self-control was he able to challenge Shingen once more, and this time slay him in their duel. Although

it took years for other writers to catch on to the change Claremont had made in the character, eventually this new characterization of Wolverine, as a man willing to kill if necessary, but in control of his capacity for violence, was universally accepted.

Claremont had one final modification to make, in *Wolverine*'s sequel, the *Kitty Pryde and Wolverine* limited series, by himself and artist Al Milgrom. Once again in Japan, Wolverine was confronted by another destructive father figure, this time his own former mentor, the allegedly immortal swordsman Ogun. Unable to best him through fighting skill, Wolverine was forced to let the beast within himself go free once again. His irrational ferocity won the final battle, but Wolverine was horrified at what he had done while in its grasp. Still, he had regained

Wolverine #57 (1992) Pencils: Marc Silvestri / Inks: Dan Green
Wolverine's love for the Japanese noblewoman Mariko Yashida ended tragically with her death. Here artists Marc Silvestri and Dan Green give a sculptural quality to the evocation of grief.

X-Men (Second Series) #25 (1993)
Script: Fabian Nicieza / Pencils: Andy Kubert / Inks: Matt Ryan
Magneto employed his magnetic powers to tear the adamantium from Wolverine's body. Only after he had done so did the readers learn that Wolverine's claws were natural parts of his skeleton, and not the adamantium constructs most of them had thought.

his rationality once the battle was over: the beast had not gone but would always be present within him, yet he could now confine it at will.

Marvel could not ignore Wolverine's tremendous popularity, and he received his own regularly published comic magazine, which continues to this day. Throughout his time writing the X-books Claremont refused to divulge Wolverine's origin, instead dropping only occasional, intriguing hints, for example, indicating that Wolverine's healing factor caused him to age slowly and that he was far older than he looked. Claremont, Byrne, and writer Bill Mantlo described how Wolverine had been found in the snowy Canadian wilderness, feral, nearly naked, incapable of speech, by government agent James Hudson and his wife, Heather, who became his surrogate family, reintroducing him to civilization. Eventually, through Hudson, he became an operative for Canadian intelligence. Later *Wolverine* writers, notably Barry Windsor-Smith and Larry Hama, finally filled out much of Wolverine's past: that he had been a special CIA operative in the 1960s and later fell victim to the secret Canadian government experiment that infused adamantium into his body.

Wolverine recently suffered a double blow: his true love, Mariko, was murdered and the archvillain Magneto drew the adamantium out from his body, rendering him vulnerable at long last to harm. And so he left the X-Men for a time, striking out on his own: the man who was once little more than an emotion-driven animal had become a stoic survivor, seeking his role in the world.

THE X-MEN IN THE EIGHTIES

The popularity of *The Uncanny X-Men* mounted throughout the 1980s, as Chris Claremont collaborated with a series of leading comics artists: among them was, first, the returning Dave Cockrum; later Paul Smith, John Romita, Jr., and Marc Silvestri; and finally, as the series entered the 1990s, Jim Lee, whose handsome and powerful drawing seized the imaginations of the latest generation of comics collectors.

Working with these and other artists, Claremont created several memorable new members for the X-Men. Perhaps the best and most popular of these was Rogue, who originally appeared as a villainess in a story Claremont did with artist Michael Golden in an *Avengers Annual*. Once she left the Brotherhood of Evil Mutants for the X-Men, Rogue's personality blossomed into that of a fun-loving, earthy southern girl in her late teens, tempering the recurring gloom of the X-Men's world with high spirits and country wit. But her facade hid her inner turmoil. Remember that physical mutations could serve as a metaphor for a person's fears about his or her own sexuality; with Rogue the connection became more explicit. Due to her mutant power, if she touched her bare flesh to that of another person, she would absorb that person's memories and special abilities, casting him or her into a coma. Usually the effects were temporary, but not always:

X-Men (Second Series) #6 (1992) Plot and art: Jim Lee / Script: Scott Lobdell / Finished art: Art Thibert

SABRETOOTH

In the pages of *Iron Fist* Claremont and Byrne had created a savage rival to Wolverine with similar abilities named Sabretooth, whom they intended to be Wolverine's father. Perhaps because so many people had guessed Sabretooth's identity, Larry Hama changed it: instead, Sabretooth was Victor Creed, Wolverine's untrustworthy former partner in the CIA. Recently Sabretooth took over the role Wolverine used to play in the X-Men: that of the potential time bomb at Xavier's mansion. But whereas Wolverine was a full member of the team, Sabretooth is Xavier's captive, kept fettered while the Professor labors to cure Creed's insanity.

Behind Sabretooth stands Psylocke, who has perhaps the most unusual history of any of the X-Men. She was introduced by Claremont in Marvel's *Captain Britain* series as Betsy Braddock, the Captain's sister, who proved to have telepathic powers. A decade later Claremont inducted her into the X-Men, and subsequently this Caucasian Englishwoman's mind was transplanted into the body of a Japanese martial artist, where it remains to this day.

Uncanny X-Men #130 (1980)
Script: Chris Claremont / Pencils:
John Byrne / Inks: Terry Austin
Created at the tail end of the disco
craze of the 1970s, the Dazzler was
a mutant who took the path
Spider-Man originally chose: to use
her powers to be an entertainer. In
the Dazzler's case, she could transform sound into light. She soon
starred in a series of her own and
later served in the X-Men; she fell
in love with fellow member
Longshot and returned with him to
his home dimension.

her other powers, her superstrength, near-invulnera-
bility, and ability to fly, were the accidental result of
permanently absorbing the powers of a former
super hero, Ms. Marvel, back in her criminal days.
Worse, Rogue was literally haunted by her victims:
traces of their personalities remained buried in her
subconscious. As a result, she was afraid of getting
close to anyone and sexuality seemed forbidden to her.

In recent years, however, she has found
romance with another Claremont creation, the Cajun

Uncanny X-Men #271 (1990)
Script: Chris Claremont / Pencils:
Jim Lee / Inks: Scott Williams
After Shadowcat moved to
England, Wolverine formed a
similarly fatherly attachment
to another mutant teenage
girl, Jubilee.

outlaw mutant Gambit, created by Claremont and
Jim Lee. A renegade member of a Thieves' Guild
based in New Orleans, Gambit uses, as his trademark weapons, playing cards that he charges with
explosive energy.

Perhaps because she had grown so much in the
role, Kitty finally left the X-Men and was eventually
replaced by another young, high-spirited teenage
girl, co-created by Claremont and Silvestri. This was
Jubilee, a Chinese-American orphan who can create
energy discharges resembling fireworks. Best
known for her unique slang and style of speaking,
which sound somehow contemporary but were
pretty much invented by Claremont, Jubilee sighted
her role models—the female X-Men—at a mall in
Southern California and decided to follow them
home. She too bonded with Wolverine, her light
balancing his dark, until he left the team and she
decided to join the newly formed team of young
mutants, Generation X.

Claremont also continued to develop long-
standing *X-Men* characters, making them his own.
Seeking to give Magneto believable motivation,
Claremont revealed him to have been a child imprisoned in the death camps at Auschwitz, a witness to

X-Factor #84 (1992) Script: Peter David / Pencils: Jae Lee / Inks: Al Milgrom

Scott Summers's brother Alex, alias Havok, and the green-haired mutant Lorna Dane, who became known as Polaris, were first introduced in *The X-Men* in the late 1960s. For many years Havok needed his special costume to control the vast amounts of cosmic energy he draws from space, which he can discharge in powerful bolts. Polaris has magnetic powers and was once deceived into thinking she was Magneto's daughter. Alex and Lorna quickly fell in love with each other and have remained so ever since. They temporarily rejoined the X-Men in the 1980s and are now members of X-Factor, where they play the same role of romantic leads that Scott and Jean long had in the original *X-Men*. (In case you were wondering, artists Jae Lee and Jim Lee are no relation to Stan, or to each other!)

Longshot #1 (1985) Script: Ann Nocenti / Pencils: Arthur Adams / Inks: Brent Anderson

The perennially cheerful adventurer Longshot was the brainchild of writer Ann Nocenti and artist Arthur Adams. Introduced in his own limited series, he served for a time in the X-Men, even though he was not a mutant Earthman. Instead, Longshot was an artificially created humanoid from another dimension, who was gifted with unusual luck.

Uncanny X-Men #173 (1983) Script: Chris Claremont / Pencils: Paul Smith / Inks: Bob Wiacek

Claremont and his collaborators kept the *X-Men*'s popularity high through his sixteen-year run on the series by continually taking into account changes in the audience's taste. Whereas Cockrum and Byrne's Storm had looked sweet, innocent, and sensitive, in the early 1980s Claremont and artist Paul Smith gave her a tougher, more contemporary image, dressing her in leather and replacing her flowing mane with a Mohawk.

Uncanny X-Men #275 (1991)
Script: Chris Claremont / Pencils:
Jim Lee / Inks: Scott Williams

AIN'T THAT JUST *TOO* BAD, HONEY-BUNCH, SURE BREAKS MY HEART.

GUESS YOU AIN'T SUCH A *HOT* LI'L NUMBER AFTER ALL.

WAH-*HOOOOO!*

AH FEEL LIKE A *MILLION*-- --AH GOT MY *POWERS* BACK!

Introduced as a forbidding-looking young villainess, Rogue (above) has grown considerably more appealing over the years in both looks and personality. Her southern earthiness and high-spiritedness make her stand out amidst many of the more somber X-Men.

Gambit uses his mutant power to imbue ordinary trading cards with explosive energies. Before joining the X-Men he traveled the world as a consummate thief, as in the flashback (below right) to an exploit of his in Paris. Because so much of his criminal past remains a mystery, stories have sometimes raised doubts as to how truly committed Gambit is to the X-Men's moral code.

In recent years an affecting romance evolved between Gambit, a man with dark secrets in his past, and Rogue, whose power to absorb minds through her touch seemed to preclude her from physical intimacy (left). Above right, they train in the X-Men's Danger Room, which is equipped with Shi'ar technology that can simulate any peril or environment, much like the "holodeck" on *Star Trek: The Next Generation.*

X-Men (Second Series) #24 (1993) Pencils: Andy Kubert / Inks: Matt Ryan

Gambit #1 (1993) Script: Howard Mackie / Pencils: Lee Weeks / Inks: Klaus Janson

X-Men (Second Series) #33 (1994) Script: Fabian Nicieza / Pencils: Andy Kubert / Inks: Matt Ryan

the scope of man's potential for hatred of his fellow man. Discovering himself to be a mutant, Magneto refused to allow the new race of Homo superior to fall victim to another holocaust. Yet his revulsion against the Nazis led Magneto to model himself inadvertently after his former tormentors, becoming the leader of another supposed master race bent on world domination. In *Uncanny X-Men* #150 Claremont forced Magneto to realize what he had become, and eventually the X-Men's greatest foe became their ally; by the premiere of the new companion *X-Men* series in the 1990s, however, Magneto, deciding that Xavier's cause was helpless, had reverted to his previous ruthless fanaticism.

Perhaps Claremont's greatest creation of his later years with the series was the nation of Genosha, which combined the dangers of uncontrolled genetic engineering, a new topic in the news, with South Africa in the closing years of apartheid. Genosha was an island, that, although located off the coast of Africa, was dominated by English-speaking whites. As in the United States, anyone could be born a mutant, but in Genosha, a person born a mutant was also born a slave. Eventually, Claremont and fellow writer Louise Simonson chronicled the fall of Genosha's apartheid, but freedom did not last long. In recent years, in the "Blood Ties" story line in *The Avengers* and *X-Men* titles, the country collapsed into civil war between its former "normal" rulers and embittered former mutant slaves, now bent on wiping out their former masters.

GENOSHA

*Uncanny X-Men #235 (1988) Script: Chris Claremont /
Pencils: Rick Leonardi / Inks: P. Craig Russell*

In Chris Claremont's initial Genosha story line, the young son of the nation's leading genetic engineer had a girlfriend who, on being discovered to be a mutant, immediately lost all her civil rights. Her head was shaved, a uniform was permanently attached to her body, and she was put to work in one of Genosha's mutant labor camps. In the course of the story Wolverine, Rogue, and the boy, who was mistaken for a mutant, all found themselves victims of this prison system. Finally, the other X-Men—Storm, Longshot, Dazzler, Colossus, and Havok—came to their rescue (left), but the boy, now radicalized, remained with his girlfriend, determined to change the system by whatever means he could.

In a subsequent story line, "The X-Tinction Agenda," members of the X-Men, X-Factor, and the New Mutants were briefly enslaved by the Genoshans. (Below, Storm wears the shaved head and skintight uniform of the country's mutant slaves.) This time the mutant heroes helped overthrow the Genoshan government, and it seemed there would at last be equality there between mutants and normal humans. But, instead, in another story line, "Blood Ties," Genosha fell into civil war between the two sides, and the chaos remains unresolved.

*Uncanny X-Men #271
(1990) Script: Chris
Claremont / Pencils: Jim
Lee / Inks: Scott Williams*

X-Factor #10 (1986) Script:
Louise Simonson / Pencils: Walter
Simonson / Inks: Bob Wiacek

The first major story line crossing through all of the growing X-Men family of titles was the "Mutant Massacre." In it a mutant team of assassins called the Marauders, then working for the X-Men's mysterious adversary Mister Sinister, slaughtered most of the Morlocks, mutant outcasts living beneath Manhattan's streets (named after a similar community in H. G. Wells's *The Time Machine*). Marvel's mutant heroes managed to save only a small percentage of the Morlocks: heroic victories were no longer a foregone conclusion in the *X-Men* books.

Here a Marauder fells a Morlock with an "energy harpoon." Note how artist Walter Simonson intensifies the horror of the murder with the victim's contorted pose, hovering in midair, and with characteristically stylized lighting effects.

By the time these comics were published, the original X-Men had dispersed and re-formed. Eventually, it was decided to reteam Xavier's five original students under a new name; this, however, meant somehow bringing Jean Grey back from the dead. The solution was to explain that she had never really died in the first place: instead, when she made that fateful flight through the solar storm, an alien entity called the Phoenix force cast her into suspended animation and then impersonated her from *X-Men* #101 onward. Once found and revived, Grey joined the other four original X-Men in forming a new team, X-Factor, in the series of the same name created by Bob Layton and Jackson Guice. The initial concept, though, was seriously flawed: the original five X-Men pretended, in their secret identities, to be normal humans who hunted and captured mutant "menaces." In secret the X-Factor members actually trained the mutants they "caught" to use their powers and pass as human in normal society. It was pointed out by critics, and finally within the series itself, that X-Factor was thus perpetuating the idea that to be a mutant—to be different—was wrong. Under the book's new husband-and-wife creative team of writer Louise Simonson and artist Walter Simonson, X-Factor publicly renounced their former modus operandi and proclaimed themselves to be mutants. The Simonsons also made their mark by transforming the Angel into a grimmer version for the 1980s: Archangel, with metallic-seeming wings he could use to fire razor-sharp feathers at his enemies.

ALPHA FLIGHT AND EXCALIBUR

Back in the 1970s, before being assigned to *The X-Men*, the British-born Claremont had, together with artist Herb Trimpe, created Captain Britain for Marvel's newborn United Kingdom line. Brian Braddock was a college science student, something like an English version of Peter Parker, who was chosen by Merlin and his daughter Roma to become the costumed champion of the British Isles, thus mixing Captain America's origin with the Arthurian mythos. Captain Britain's adventures continued off and on over the following decade, chronicled by British writers and artists, including early work by writer Alan Moore, who would revolutionize comics writing in the 1980s with *Swamp Thing* and *Watchmen* at DC, and artist Alan Davis. In the late 1980s Claremont and Davis made Captain Britain the center of a new British-based team of super heroes, Excalibur (right). Besides the Captain (second from left) and his lover, the shape-shifting mutant Meggan (far left), the team also included the X-Men's Rachel Summers, the second Phoenix (center); Kitty Pryde, now known as Shadowcat (second from right); and Nightcrawler (far right). There was a sense of whimsy to Excalibur's exploits that distinguished their series from the other X titles and enabled it to incorporate everything from comedic villains like Arcade and the Crazy Gang to a family of talking dinosaurs exploring London while garishly dressed as stereotypical American tourists.

Excalibur #1 (1988) Pencils: Alan Davis / Inks: Paul Neary

During his term on *Uncanny X-Men*, John Byrne created his own super hero team, Alpha Flight, an official arm of the Canadian government headed by James MacDonald Hudson, Wolverine's old friend and benefactor. Later, Byrne started writing and drawing an Alpha Flight series, and although he left after only a few years, *Alpha Flight* continued for 120 issues.

Looming behind the others in this group shot (below left) is Sasquatch, a human scientist who can transform into the legendary Canadian beast. The others, from left to right, are Northstar, a French-Canadian who can move at superhuman speed; Snowbird, who was the daughter of an Inuit goddess and could change into any Arctic animal; the Native American mystic Shaman; the acrobatic dwarf Puck; James Hudson, known at different times as Guardian and Vindicator; Marrina, an alien who could breathe in and out of water; and Northstar's sister, Aurora, who shares his powers. After Hudson was believed to have died, his wife Heather assumed leadership of the team as well as his costumed Guardian persona; when he returned they worked as partners.

Aurora is a prime example of how a hero's dual identity can represent a fragmented psyche. In her "normal" identity of Jeanne-Marie Beaubier, she is a staid, repressed Catholic school teacher, but as Aurora she is sensual and even exhibitionistic. As for her brother, Northstar, Byrne hinted strongly almost from the beginning that Northstar was gay, making him Marvel's first homosexual super hero; years later, *Alpha Flight* writer Scott Lobdell finally had Northstar clearly declare his sexual orientation.

House advertisement for *Alpha Flight* (1984) Art: John Byrne

THE NEXT GENERATION

Once *The Uncanny X-Men* had become Marvel's top-selling book, it was inevitable that there would be a spin-off title, featuring a new group of mutants. And that is exactly what the team was called: the New Mutants, created by Claremont and artist Bob McLeod, made their debut in Marvel's fourth graphic novel, published in 1982, and moved into their own regularly published comic almost immediately thereafter, which was soon to feature the extraordinary art of Bill Sienkiewicz.

Despite the presence of Kitty Pryde and the obligatory Danger Room training sessions, the X-Men of the 1980s could no longer truly be considered a school; even the next youngest member, Colossus, was over voting age. As the comics readership included increasingly larger numbers of teenagers and adults, perhaps the fact that the X-Men were themselves now mostly adults increased their appeal. "Professor Xavier's School for Gifted Youngsters" was now more like a commune inhabited by a clique of friends who had mastered their skills and merely needed to keep honing them through practice, as if the Danger Room were a more stressful version of a health club.

Still, there was a feeling at Marvel that the idea of a school for mutants was essential to the original concept behind the X-Men, and so the solution was to give Xavier a new class to teach. Thinking (not for the first time) that the X-Men had perished in battle, Xavier recruited a new band of mutant youngsters. Except for the Vietnamese Karma, they were all adolescents: Cannonball, from a mining family down South, who could propel himself through the air like a human cannonball without the cannon; Sunspot, a Brazilian whose hot temper was appropriate for someone whose strength was amplified by solar energy; Wolfsbane, a sexually repressed Scots orphan who found herself transforming into a werewolf, with animalistic instincts to match; and Mirage, a Native American who could manifest mental images of the object of a person's greatest desire or fear.

Two of the most popular members of the New Mutants did not quite fit the series concept: one was not a human mutant, and the other was something more. The latter, Illyana Rasputin, called Magik, Colossus's younger sister, had mutant teleportational abilities, but her main power was sorcery. In his *Magik* limited series, drawn by John Buscema, Claremont wove a dark fairy tale, reworking his recurring theme about the potential for evil in the soul of the innocent. It recounted how the prepubescent Illyana was kidnapped from Earth by the demonic sorcerer Belasco, who forced her to serve as his apprentice in black magic, thereby corrupting her soul. After spending seven years in his other-dimensional limbo mastering sorcery, she bested Belasco in mystical combat and drove him from his

New Mutants #19 (1984)
Script: Chris Claremont /
Art: Bill Sienkiewicz
Like Stan Lee, Chris Claremont succeeds in portraying troubled adolescents in whom the readers can see their own anxieties reflected. From left to right are the New Mutants Cannonball, Wolfsbane, Magma (a later addition to the team who can summon up volcanic lava), and Magik. Sienkiewicz's lighting of Magik hints at the dark mystical forces within her soul.

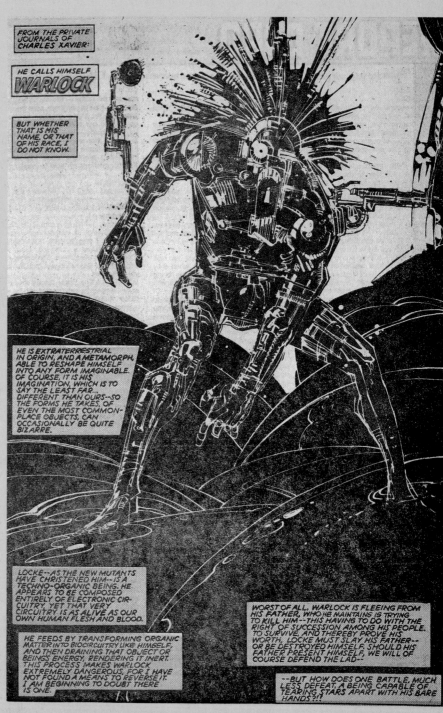

New Mutants #19 (1984) Script: Chris Claremont / Art: Bill Sienkiewicz

New Mutants #23 (1985) Script: Chris Claremont / Art: Bill Sienkiewicz

Artist Bill Sienkiewicz stayed on *The New Mutants* for only two years, but he helped make the series one of the most artistically daring comics of the 1980s. He could depict the New Mutants in photo-realist styles, but he was continually experimenting, shattering the accepted, supposedly realistic conventions for depicting the world of the Marvel super heroes. Warlock (right), for example, a machinelike "techno-organic" alien, looked like nothing seen in the Marvel Universe before. It was not simply that he was an alien: the very style in which he was drawn, influenced by the work of illustrator Ralph Steadman, set him apart from the more conventionally depicted world and people around him. Magik was also a perfect subject for Sienkiewicz, since her supernatural powers could be depicted in novel and exiting ways. Left, Magik and Sunspot battle a demonic bear that haunted their teammate Mirage. *The New Mutants* was a key book for Sienkiewicz's stylistic transition from the Neal Adams-influenced *Moon Knight* to the surreal eclecticism of the forthcoming *Elektra: Assassin*.

X-FORCE / ROB LIEFELD

Under Rob Liefeld as co-plotter and penciller, the charming and relatively innocent team of students that were the New Mutants quickly evolved into the hard-edged band of warriors known as X-Force. Pictured here, from left to right, are the Boomer, former new Mutant Cannonball, the lupine Feral (kneeling), X-Force's mentor Cable, the superhumanly strong Native American Warpath, Cable's longtime partner Domino (kneeling), and the other-dimensional swordsman Shatterstar.

With his large panels of massively muscled figures toting enormous guns in dynamic poses, Liefeld commanded tremendous enthusiasm from a new generation of comics fans and rapidly spawned a host of imitators at many comics companies. Liefeld, the *X-Men*'s Jim Lee, and *Spider-Man*'s Todd McFarlane were key figures in radically transforming the look of super hero comics. There was no mistaking a "hot" comic of the early 1990s for one of the previous decade.

Pencils: Rob Liefeld / Inks: Brad Vancata

Deadpool: The Circle Chase #3 (1993)
Script: Fabian Nicieza / Pencils: Joe
Madureira / Inks: Harry Candelario

It speaks volumes about the 1990s that comics villains with sufficient charisma kept ending up as the stars of their own books. This was the case with the seemingly indestructible masked assassin Deadpool, whom Rob Liefeld had introduced into *The New Mutants* as their antagonist. Perhaps his popularity was due to the nonstop free-associating banter that scripter Fabian Nicieza gave him, making Deadpool much like Spider-Man gone bad. Like Venom, once Deadpool got his own limited series, he started to develop something of a conscience, though he did not mellow by much.

own realm. Illyana then returned to Earth to find that she had been gone only a few seconds, despite having aged seven years. Back at Xavier's mansion she was a schoolgirl, but she was also princess of a dimension of demons, with a darkness to her soul that frightened the unwary who looked into her eyes.

The other unlikely New Mutant was actually an alien named Warlock. His origin was a Freudian myth in a comedic mode. Warlock was a "techno-organic" alien child, a living, shape-shifting machine. But he was also a prince of an alien world, where the ruler, his father, the Magus, slew his offspring lest they become greater than he and depose him. Playful and innocent and hardly a warrior, Warlock fled through outer space, crash-landing on Earth near Xavier's school, where he found the loving surrogate family he lacked.

Although the New Mutants proved the equals of adult adversaries as fighters, they were school kids with their own school uniforms: maskless versions of the original X-Men costumes. They visited the local mall, held slumber parties, and even went to a school dance at the alma mater of their rivals, the teenage mutant Hellions, who were trained by the Hellfire Club's White Queen! When Xavier went on a mission into outer space, he left the school in the care of a temporarily reformed Magneto, who thereupon played the unexpected role of a brooding but benign headmaster, continually and vainly trying to restrict the New Mutants to the school grounds after their latest misadventure.

Perhaps the boarding-school atmosphere of the New Mutants was somewhat too innocent and naive for the young readers of the 1990s. Under Bob Harras's editorship, writer Louise Simonson and artist Rob Liefeld began reworking the team into a band of outlaw commandos battling mutant terrorists. On Harras's suggestion that they create a new leader for the New Mutants, Liefeld and Simonson came up with a character who seemed as thoroughly different from Charles Xavier as possible. The man known as Cable was a cyborg, part-man and part-machine, and an outlaw. He thought Xavier's ideal of coexistence between mutants and "normal" people was a naive fantasy. There was a war coming, he believed, and he was determined to train the New Mutants to fight in it like soldiers.

Eventually, Cable's attitude led to *The New Mutants'* being supplanted by a new series, plotted and drawn by Liefeld and scripted by the rising star Fabian Nicieza, featuring the revamped team as *X-Force*. Cannonball and Sunspot were joined by Boomer, who had first been introduced as a teenage Madonna wannabe who could project explosive energy balls; the Latino Rictor, who, as his name implied, could trigger small earthquakes; Banshee's daughter Siryn, who had powers similar to her father's; the late Thunderbird's brother Warpath; and the other-dimensional warrior from the future Shatterstar, who would prove to be Longshot's son.

THE ADVENTURES OF CYCLOPS AND PHOENIX

duplicate, whom Apocalypse, mistaking him for the original, kidnapped. Transported to this future era, Scott Summers and Jean Grey took on the guises of "Slym" and "Redd" and raised Nathan in the wilderness (below), training him in the use of his powers.

Meanwhile Apocalypse raised Nathan's evil twin—Stryfe—as an Antichrist, an heir who would perpetuate his tyranny. Young Nathan proved instrumental in overthrowing Apocalypse, and Scott and Jean were drawn back to their own time. Nathan, however, remained behind; his further adventures as a youth are recounted in another limited series by Lobdell, scripter Jeph Loeb, and artist Gene Ha, *Askani'son*. Eventually he matured into Cable and became the archenemy of Stryfe. Completing the circle, Cable traveled back to the twentieth century, knowing that it was the crucial period in the history of Homo superior, and there reshaped the New Mutants into X-Force, as readers had already seen. (By the series' end, Jean Grey adopted the name Phoenix, in honor of Rachel, who had by then died.)

The Adventures of Cyclops and Phoenix #1 (1994) Script:
Scott Lobdell / Pencils: Gene Ha / Inks: Al Vey

After Rob Liefeld's departure from *X-Force*, writer Fabian Nicieza developed Cable into a more emotionally balanced and mature character. It turned out that there was far more to Cable than there seemed when he first appeared. Cable proved to be the son of Scott Summers, a.k.a. Cyclops, and the late Madelyne Pryor, a Jean Grey look-alike whom Scott had married and who later proved to be Jean's clone. The child, named Nathan, contracted the strange "transmode virus," which turns flesh into "organic circuitry" (above) and would have died had Scott not turned him over to an enigmatic messenger from a far future time.

Nathan's fate was revealed in author Scott Lobdell's limited series *The Adventures of Cyclops and Phoenix*, a futuristic variation of the classic mythic story line of the birth and youth of the hero destined to redeem a world filled with evil. Scott Summers proves to be the progenitor of a bloodline destined to change humanity's future. The envoy from the future who took the infant Nathan from Cyclops was a member of the Askani, an underground resistance movement two millennia hence, when humanity was ruled by Apocalypse. The Askani were led by the aged Rachel Summers, who had been known in the X-Men's own time as the second Phoenix. They halted the spread of the transmode virus in the baby's body, saving his life but leaving him frail, and they also cloned the child, creating a healthy

The Adventures of Cyclops and Phoenix #2 (1994) Script:
Scott Lobdell / Pencils: Gene Ha / Inks: Al Vey

X-Men (Second Series) #1 (1991) Pencils: Jim Lee / Inks: Scott Williams
The X-Men circa 1991, from left to right: Jean Grey, Professor X, Storm, the Beast, Archangel (the new name taken by the Angel after he was transformed by Apocalypse), Colossus, Gambit, Rogue, Psylocke, Cyclops, Wolverine, and Iceman. On the far right is Magneto.

THE X-MEN IN THE NINETIES

At the beginning of the 1990s, the original five X-Men returned to the pages of *Uncanny X-Men*, preparatory to the series splitting into two: *Uncanny* and the new companion series, titled simply *X-Men*. *X-Factor* became a wholly different book, featuring a new team of mutants working as operatives of the federal government, as initially scripted by Peter David, a writer known for his ironic subversion of the conventions of the super hero genre. The new members included former X-Men Havok and Polaris, Quicksilver, Wolfsbane, and Strong Guy, whose name was David's apt commentary on the essence of super hero names.

Despite the transformation of *The New Mutants* into *X-Force*, the concept of a school for mutants was far from finished. It appeared that every decade required the recruitment of a new class for Xavier. Hence in 1994 came the debut of the fourth "generation" of students, aptly named *Generation X*, created by writer Scott Lobdell and artist Chris Bachalo. This time Xavier turned over the job of teaching to the longtime X-Man Banshee and, surprisingly, the White Queen, who had reformed, at least for the time being. In creating members for

X-Factor #101 (1994) Pencils: Jan Duursema / Inks: Allen Milgrom
Whereas the original X-Factor was a tight-knit surrogate family (namely, the original X-Men), the members of the current X-Factor have very little in common apart from being mutants. The exceptions are the longtime couple Havok (in the center) and Polaris (levitating in front of the logo). The enormous figure on the left is Strong Guy, who, unsurprisingly, was first drawn by Bill Sienkiewicz: Strong Guy absorbs energy and converts it into strength, thereby distorting his musculature to grotesquely huge proportions. Former New Mutant Wolfsbane crouches at the bottom. On the right is a portrait of Forge, a Native American mystic whom mutation gave an intuitive genius for invention. Smoking a cigar is X-Factor's former ally Random, a mercenary who can alter parts of his body into weapons.

GENERATION X

MUY BUENO, MU-CHACHO.

THANK YOU FOR ASKING.

WHEN THE REST OF THE KIDS ON THE BLOCK STARTED GROWING *FACIAL HAIR* AND FOUND THEIR *VOICES CHANGING*...

...ANGELO ESPINOSA DISCOVERED HE WAS GROWING ABOUT SIX FEET MORE SKIN THAN HE NEEDED.

HE'S ADJUSTING.

Generation X #1 (1994) Script: Scott Lobdell / Pencils: Chris Bachalo / Inks: Mark Buckingham

The members of Generation X (below) and their faculty, from left to right: the Banshee, who acts as teacher; M, an Algerian with superhuman strength and near invulnerability; Jubilee, formerly of the X-Men; Husk, who is the sister of X-Force's Cannonball and can change form by shedding her outer skin; Skin (also pictured left), a Latino from Los Angeles's inner city; Mondo, a Samoan "omnimorph" who can take on the properties of any form of matter; Chamber, whose energy powers destroyed half his face; another teacher, the White Queen; Penance, with her razor-sharp skin; and Synch, who can tap into other mutants' energies.

Generation X Collectors' Preview #1 (1994) Art: Chris Bachalo

Uncanny X-Men #277 (1991) Script: Chris Claremont / Pencils: Jim Lee / Inks: Scott Williams

JIM LEE

Jim Lee had tremendous impact on both comics readers and other super hero artists of his generation. His talent at drawing handsome, heroically proportioned figures and vivid spectacles is displayed in this double-page spread of X-Men Forge (on the left, in the background), Banshee (center), and Storm (on the right) in outer space.

their new team Lobdell and Bachalo devised new metaphors for the physical and emotional tensions and fears of adolescence. There was Skin, who has several feet more skin than his body requires; Husk, who can shed her skin at will, revealing a layer underneath that will not necessarily look human; and Chamber, who discovered his ability to project energy bolts when they accidentally ravaged part of his own body.

The first issue of the new *X-Men* comic book in 1991, a companion to the original *Uncanny* series, sold eight million copies, an all-time record in the comics industry. Chris Claremont ended his long reign as *X-Men* writer by teaming up with artist Jim Lee for the new book's first three issues, which marked Magneto's return to his traditional role as

the X-Men's primary nemesis. Lee and fellow artist Whilce Portacio plotted and drew *X-Men* and *Uncanny* for roughly another year; with their departures for new projects the writing reins fell into the hands of Fabian Nicieza and Scott Lobdell, who skillfully made the books their own, along with such artists as Andy Kubert and Joe Madureira.

The traditional themes of the X-books continued to prove capable of finding new expressions with each decade. Portacio introduced Bishop, another mutant time traveler, in this case a member of a twenty-first-century mutant police force that idolized the X-Men of the previous century. Yet another effective metaphor was devised linking the fate of Marvel's mutants to that of another minority: as AIDS ravaged the real world Xavier and his X-Men

X-Factor #24 (1988) Script: Louise Simonson / Pencils: Walter Simonson / Inks: Bob Wiacek

APOCALYPSE AND STRYFE

First appearing in *X-Factor*, the mutant shapechanger Apocalypse now even threatens to unseat Magneto as the X-Men's preeminent foe. Having lived for thousands of years, Apocalypse (left) was worshiped as the death god of various past cultures; he is, in effect, an ancient deity recast in science fictional terms, attempting to crush the contemporary world. The ultimate Social Darwinist, Apocalypse, in contrast to Xavier, seeks to foment warfare between mutants and normal humans so that only those who prove to be the strongest—including, of course, himself—will survive. In one alternate future Apocalypse will live on for two more millennia, transferring his mind from one body into another, finally conquering Earth only to meet his doom at the hands of the young Cable.

The terrorist Stryfe (below), raised by Apocalypse, who encouraged his viciousness, is Cable's literal double and evil counterpart. Their personal war was chronicled in "The X-Cutioner's Song." Posing as Cable, Stryfe shot and nearly killed Xavier; then, in his own guise, he had a showdown with his future foster father, Apocalypse. Finally, locked in hand-to-hand combat with Cable, Stryfe was physically annihilated, but his consciousness infused itself into Cable's, literally becoming the shadow side of Cable's psyche. For a time Stryfe's persona took charge, but Cable succeeded in regaining control.

X-Force #17 (1992) Script: Fabian Nicieza / Pencils: Greg Capullo / Inks: Harry Candelario

245

THE AGE OF APOCALYPSE

In 1995 *X-Men* editor Bob Harras and writers including Warren Ellis, Larry Hama, Scott Lobdell, and Mark Waid undertook a reworking of the themes of Claremont and Byrne's "Days of Future Past" on a massive scale. For three full months, every *X-Men*–related series was taken over by the story line known as "The Age of Apocalypse," which temporarily transformed not only the Marvel Universe but also the titles of the regularly published *X-Men* line of comics (opposite).

It began when Xavier's illegitimate son Legion traveled back in time to a point before the founding of the X-Men, when Xavier and Erik Lehnsherr, the future Magneto, were friends, albeit with very different views on how mutants could win equal rights in human society. Legion intended to kill Magneto, thus changing history and sparing Xavier all the misery Magneto had caused him over the succeeding years. But Xavier intercepted Legion's lethal psionic blast and was killed before Lehnsherr's eyes.

Weapon X #2 (1995) Script: Larry Hama / Pencils: Adam Kubert / Inks: Dan Green

The startling result was the creation of a new alternate time line, "The Age of Apocalypse," in which Xavier had never led the X-Men against the mutant menaces that arose over the decades. Instead, Magneto, shocked by his friend's sacrifice, vowed to devote his life to Xavier's dream of peaceful coexistence between normal humans and mutants. Hence, it was Magneto who founded the X-Men; he even married Rogue and they had a son, whom he named Charles.

But Magneto was apparently not the inspirational leader that Charles Xavier would have been. He could not prevent Apocalypse and his mutant cadre from taking over all of North America, slaughtering the human population and herding the survivors into concentration camps. People fleeing Apocalypse's tyranny found refuge in Europe, as war loomed between the two continents. (Left: an armada of flying Sentinels transports human refugees from North America, dominated by Apocalypse's mutant tyranny, to United Europe, which still remains under human control. The Jean Grey of this alternate time line can be seen holding refugee children, while this reality's Logan calls to her from the background.)

Much of the interest of "The Age of Apocalypse" lay in the treatment of familiar characters in this alternate reality. Some had radically changed: Havok was now a neo-fascist in Apocalypse's service, and the Beast had become a latter-day Dr. Mengele, subjecting human captives to perverse experiments. There was no Cable, but instead there was someone very like him, Nate, alias X-Man, a hot-tempered youth who could barely restrain his titanic psionic powers. Yet others remained at heart the same: though Cyclops too served Apocalypse, he finally openly rebelled against him; he and former X-Man Jean Grey found one another and fell in love in this reality too.

In the end Magneto's X-Men and their allies overthrew Apocalypse just before their reality seemingly was destroyed. History was set right, and the normal continuity resumed, in which Professor Xavier was alive and leading the X-Men. Still, some characters from "The Age of Apocalypse" somehow crossed into the normal time line, most notably X-Man, whose exploits continued in his own series.

▶ *Weapon X* #1 (1995)
Art: Adam Kubert

Astonishing X-Men #1 (1995)
Pencils: Joe Madureira / Inks: Tim Townsend

▶ *Generation Next* #1 (1995)
Pencils: Chris Bachalo / Inks: Mark Buckingham

Gambit and the X-Ternals #1 (1995) Pencils: Tony Daniel / Inks: Kevin Conrad

▶ *Factor-X* #1 (1995)
Pencils: Steve Epting / Inks: Al Milgrom

X-Man #1 (1995)
Pencils: Steve Skroce / Inks: Bud LaRosa

Uncanny X-Men #282
(1991) Pencils: Whilce Portacio / Inks: Art Thibert In an alternate future, Bishop was a member of the XSE, a mutant police force inspired by the history of the X-Men. Bishop's own mutant power was the ability to absorb kinetic energy used against him and redirect it at will. Journeying to the late twentieth century in pursuit of a time-traveling mutant criminal, Bishop found himself stranded there and joined the X-Men. For him it was like an American of today traveling back to 1776 to meet the Founding Fathers.

struggled to find a cure for the Legacy Virus, a disease that struck first at mutants and later spread to other human beings.

Much of the sustaining power of the *X-Men* over more than thirty years derives from its creative teams' loyalty to its underlying concept: they have continued to find new ways of dramatizing for each new generation of readers the concept of mutation, of what it means to be different from others.

Three characters in particular have remained central to the success of the *X-Men* from its first issue right through the present. One is the young mutants' middle-aged mentor, Professor Charles Xavier. There is an affecting irony in his very appearance: said to be the most powerful telepath on Earth, yet physically confined to his wheelchair. Oddly enough, although Chris Claremont filled out a great deal of Xavier's past, giving him former lovers and even an illegitimate son, he used the professor surprisingly little over his long term as *X-Men*

THIS SASH SERVES AS A SYMBOL OF THE MATRIMONIAL TIES.

THE SACRED PACT YOU MAKE BEFORE GOD, AN EXCLAMATION OF LOVE, WHICH WILL SERVE YOU THROUGH GOOD TIMES AND BAD.

THE WEDDING VOWS TO BE SPOKEN HERE TODAY, HAVE BEEN PREPARED BY THE BRIDE AND GROOM THEMSELVES.

"THERE WERE TIMES I WAS LOST, AND YOU FOUND ME.

"THERE WERE DAYS WHICH WERE HEAVY, AND YOU LIGHTENED MY HEART.

"THROUGH IT ALL, SINCE THE DAY WHEN WE MET, THERE WAS YOU FOR ME AND ME FOR YOU.

"THAT HASN'T CHANGED. THAT WILL NEVER CHANGE.

"TIMES HAVE BEEN GOOD, AND TIMES HAVE BEEN BAD, AND STILL, OUR LOVE HAS ENDURED AND TRIUMPHED--

MOST OF THE GUESTS HAVE LEFT. AND WHAT IF THEY DID KNOW ABOUT US, CHARLES?

THAT I'M A MUTANT?

OR THAT I'M A WOMAN WHO WANTS TO CELEBRATE THE BEST DAY OF HER LIFE?

IF DANCING WITH YOU IS WHAT MAKES ME HAPPY--

--AND IF BEING A MUTANT GIVES ME THE OPPORTUNITY TO DO THAT--

X-Men (Second Series) #30 (1994)
Script: Fabian Nicieza / Pencils: Andy Kubert / Inks: Matt Ryan
Scott Summers and Jean Grey are married before an audience of family, friends, and teammates on the grounds of Xavier's mansion, the longtime home of the X-Men (left). Aided by her levitational abilities, Professor Xavier dances with Jean at the reception (above).

writer; perhaps Xavier no longer seemed necessary to a team now composed mostly of adults. Claremont even dispatched him into outer space, traveling with the interstellar freedom fighters called the Starjammers, for many years. Xavier returned in the 1990s, and, although he principally acts behind the scenes, he has taken on greater symbolic importance to the series. Xavier is no longer simply the teacher he was in the early years of *The X-Men*; indeed, in the new *Generation X* series he has ceded that function to others. Xavier is now instead treated

as a visionary, the man who has propounded a dream of peaceful coexistence between mutants and the rest of humanity. Indeed, as the vision of prejudice in the X-books has grown darker over recent years, as even heroic characters like Cable challenge the practicality of Xavier's principles of tolerance, the books' emphasis on the necessity of implementing Xavier's vision has only grown stronger. In effect Charles Xavier is the spiritual and philosophical leader of the mutants' civil rights movement. An entire story line, "The X-Cutioner's Song," took as its basis the mutant terrorist Stryfe's attempt to destroy that vision by assassinating its architect, an attempt that inevitably recalls the political assassinations of recent decades. But Xavier survived the shooting as he survived his crippling injuries in the past: the man's ability to survive and endure seems to symbolize the persistence of his ideals, a secular faith by which all his mutant "families"—the X-Men and others—are united.

Then there are Scott Summers and Jean Grey. Issue #30 of the new *X-Men* series, published in 1994, marked the thirtieth anniversary of the concept by at long last celebrating their wedding. In the 1960s they were too introverted to speak of their feelings for one another; in 1980 their love seemed doomed by the death of Dark Phoenix. Yet though the Marvel Universe has allegedly become a darker, grimmer place in recent years, its hold on readers' imaginations ultimately lies in the depth and appeal of its flagship characters. Through all of the misfortunes that the Marvel Universe could throw at them

(and all the twists and turns with which its editors and writers shaped their fates), the romance of Scott and Jean somehow endured and triumphed.

Xavier is the mind and spirit of the X-Men, and Scott and Jean are the team's heart. Their wedding, before an audience composed of the members of the X-Men and all Marvel's other mutant teams, was Xavier's moment as well. At the reception, when Jean danced with the professor (using her powers to free him from his wheelchair) toward the issue's conclusion, Xavier's vision of a world of tolerance and harmony, of peace and joy for all people, no matter what their differences, for a moment became not just a visionary's fantasy but a dream come true.

X-MEN 2099

Even in the alternate future America depicted in Marvel's 2099 line of titles, the persecution of mutants persists. Hence a new team of mutants banded together under a time-honored name, becoming the X-Men of 2099 in the series created by writer John Francis Moore and penciller Ron Lim. Eventually this team settled with other mutants in their own community, Halo City, built in the American West under the supervision of the Doctor Doom of 2099. Pictured from left to right are Bloodhawk, Krystalin, Desert Ghost, Skullfire, Metalhead, Cerebra, and Meanstreak.

Promotional art (1993) Pencils: Ron Lim / Inks: Adam Kubert

AFTERWORD

As long as the Marvel characters appear in new stories, they will continue to evolve. Hence, even though this book can tell you what has happened to the Marvel heroes up to the time it goes to the printer in April 1996, when you read this, their lives will have taken new turns. For example, Spider-Man and Peter Parker will once again be one and the same, and Nick Fury will turn up alive before the end of 1996.

Already in early 1996 there have been startling changes in the Marvel Universe. Incredibly, it was revealed that Kang the Conqueror had subtly been manipulating Tony Stark's mind for years, finally making him his slave. Iron Man fought the Avengers and finally perished after breaking free of Kang's control. Members of the Avengers had already gone back in time and brought a nineteen-year-old Tony Stark (presumably from an alternate timeline) to their present, and he has become the new Iron Man.

In the past there had been occasional team-ups between Marvel characters and DC super heroes. But not until 1996 was there a crossover *en masse*, involving the major members of both of the two great pantheons of super heroes. The *Marvel Versus DC* limited series (titled *DC Versus Marvel* in alternate issues), coordinated by Marvel editor Mark Gruenwald and DC editor Mike Carlin, focused on pitting DC and Marvel heroes in battle against each other. Readers voted by mail and online to decide the winners of the principal matches. By popular vote, Superman overpowered the Hulk and Batman narrowly defeated Captain America. However, by the readers' decree, Spider-Man overcame DC's new Superboy, Storm bested Wonder Woman, and Wolverine trounced DC's Lobo.

At the end of the third issue of this crossover series, the Marvel and DC universes merged into one, and for a week the two companies published "Amalgam" comics in which a DC hero and Marvel hero had likewise merged into one, or a hero from one universe took on the persona of another, as in *Super-Soldier*, in which Captain America gained Superman's powers. Just as normal Marvel continuity resumed after "The Age of Apocalypse" concluded, so too in *DC Versus Marvel* #4, the Amalgam Universe split back into the normal Marvel and DC universes.

The most surprising development of all came in the summer of 1996 when a new villain, Onslaught, engaged in a tremendous battle with the Marvel heroes. As a result, the Fantastic Four, many of the Avengers including Captain America, Iron Man, and Thor, and various related characters vanished from the Marvel Universe. These characters then re-appeared in another "pocket" universe, in a new timeline, in which they began their histories over again, and their lives took different paths.

What happened was that Marvel relaunched four of its core titles, starting their continuity over again from scratch: Jim Lee now oversaw *The Fantastic Four* and *Iron Man*, while Rob Liefeld supervised *The Avengers* and *Captain America*. For the next year, beginning in August 1996, readers will see new versions of these characters' origins and new versions of their first encounters with characters such as Doctor Doom and Galactus. Lee and Liefeld have set themselves the monumental challenge of trying to match the brilliance of the classic original stories by Stan Lee and Jack Kirby.

Meanwhile, the rest of the Marvel Universe goes on as before, with the characters left in the original universe, such as Spider-Man and the X-Men, wondering what has become of their missing colleagues.

At this point Lee and Liefeld are committed to doing these four books only for a year. It is not yet known whether or not the characters re-created in the "pocket" universe will return to the Marvel Universe and resume their past histories at the end of that time. In either case it appears that Marvel has reached a watershed. Up until now the consistency of its history and fictional cosmos has been paramount; Marvel's new willingness to revise that history and even to split its universe in two presages radical new ways of developing the characters that cannot yet be predicted.

No matter how the characters change over time, however, the power of their underlying concepts will remain unaltered. The Marvel Universe is a parallel fictional cosmos into which we can project our hopes, fears, and ideals. For children these tales of heroic fantasy serve as modern fairy tales, but through increasing the sophistication of their work, Marvel's best writers and artists extended the genre's appeal to adults. The heroes are our avatars, projections of the better sides of ourselves, struggling to rise above their frailties and faults, as we do with our own. They battle injustices and oppressors just as we contend against the forces that weigh down upon our own lives. Their sufferings provide us with a catharsis; their triumphs give us inspiration. In their best work Marvel's writers and artists have created late-twentieth-century myths of humanity's innate heroism that are sure to endure long into the twenty-first.

Promotional art (1996)
Captain America
Pencils: Rob Liefeld
Iron Man
Pencils: Jim Lee / Inks: Scott Williams

Rob Liefeld and Jim Lee rework Iron Man and Captain America in their own styles.

INDEX

REFERENCE USE

ACKNOWLEDGMENTS

I would like to thank the following people: Mark Gruenwald, Michael Hobson, Julia Molino, and Lara Stein for launching me on this project and supporting my efforts; Mike Thomas, Steve Behling, and John Conroy for providing artwork and creative input; Richard Howell, Jim Krueger, Nicholas Himmel, and Richard Slovak for graciously lending vintage comics; Roger and Rafael of Roger's Comics in New York and Bill Daniels of Mile High Comics in Denver for their assistance in locating other hard-to-find comics; Eric Himmel, my editor, for his never-failing encouragement and patience; and Dana Sloan for her design.